Adaptive Radar Signal Processing

BICENTENNIAL 1807 WILEY 2007 BICENTENNIAL

THE WILEY BICENTENNIAL–KNOWLEDGE FOR GENERATIONS

Each generation has its unique needs and aspirations. When Charles Wiley first opened his small printing shop in lower Manhattan in 1807, it was a generation of boundless potential searching for an identity. And we were there, helping to define a new American literary tradition. Over half a century later, in the midst of the Second Industrial Revolution, it was a generation focused on building the future. Once again, we were there, supplying the critical scientific, technical, and engineering knowledge that helped frame the world. Throughout the 20th Century, and into the new millennium, nations began to reach out beyond their own borders and a new international community was born. Wiley was there, expanding its operations around the world to enable a global exchange of ideas, opinions, and know-how.

For 200 years, Wiley has been an integral part of each generation's journey, enabling the flow of information and understanding necessary to meet their needs and fulfill their aspirations. Today, bold new technologies are changing the way we live and learn. Wiley will be there, providing you the must-have knowledge you need to imagine new worlds, new possibilities, and new opportunities.

Generations come and go, but you can always count on Wiley to provide you the knowledge you need, when and where you need it!

WILLIAM J. PESCE
PRESIDENT AND CHIEF EXECUTIVE OFFICER

PETER BOOTH WILEY
CHAIRMAN OF THE BOARD

Adaptive Radar Signal Processing

Edited by

Simon Haykin
McMaster University
Hamilton, Ontario, Canada

WILEY-INTERSCIENCE
A John Wiley & Sons, Inc., Publication

Published by John Wiley & Sons, Inc., Hoboken, New Jersey
Published simultaneously in Canada

Library of Congress Cataloging-in-Publication Data

Adaptive radar signal processing / edited by Simon Heykin.
p. cm.
"A Wiley-Interscience publication."
Includes bibliographical references and index.
ISBN-13: 978-0-471-73582-3
ISBN-10: 0-471-73582-5
1. Rader. 2. Adaptive signal processing. I. Haykin, Simon S., 1931–

TK6580.A35 2006
621.3848–dc22

2006045743

Printed in Singapore
M WEP291353 190624

This Book
is dedicated to the memory of
Henry Booker
for his contributions to Radio Science.

Contents

Preface

For over 20 years, spanning the 1980s, the 1990s, and the early 2000s, I committed much of my research effort to two radar signal-processing applications:

1. **The angle-of-arrival estimation problem in the presence of multipath**, which is exemplified by a low-angle radar designed to track a sea-skimming missile.
2. **The reliable detection of a small target in the presence of sea clutter** (i.e., radar backscatter from an ocean surface); such a target could represent a fishing boat or a small piece of ice broken away from an iceberg floating in the ocean.

Both of these problems pertain to a marine radar environment, hence the decision to put them as integral parts of this book; moreover, they do share some common signal-processing considerations.

Equally important is the fact that both problems are challenging in both theoretical as well as practical terms.

Except for the introductory chapter 1, each of the remaining five chapters starts with introductory remarks, concludes with discussion, and ends with a comprehensive list of references of its own.[1] Each chapter is essentially self-contained, cross-references between chapters are made wherever appropriate. Moreover, the Discussion not only summarizes the important findings reported in a particular chapter, but also looks beyond those findings, encouraging the pursuit of further research.

Acknowledgments

The writing of this book has been made possible by the research contributions of many graduate students, post-doctoral fellows, and research colleagues, with whom it has been a pleasure to work over the years. In particular, I would like to express my deep gratitude to the following contributors:

- Anastasios Drosopoulos for the theoretical and experimental work done on the angle-of-arrival estimation problem, as part of his Ph.D. thesis.
- Vytas Kezys and Edward Vertatschitsch for building the MARS research facility.

[1] Except for Chapter 2, the references are listed in the order in which they are cited in the text. In Chapter 2, following the original article on which the chapter material is based, the references are listed in alphabetical order.

- Tarun Bhattacharya for the work he did on designing a neural network-based receiver for the coherent detection of a weak target in clutter.
- David Thomson for pioneering the multi-taper method (also known as the multiple-window method).
- Brian Currie who spent more than 25 years working with me as a research collaborator on numerous radar projects.
- Rembrandt Bakker for significant contributions to sea clutter dynamics and Bayesian target detection.
- Maria Greco and Fulvio Gini for extending our work on the hybrid amplitude modulation/frequency modulation model of sea clutter by accounting for nonstationarity of the clutter.
- Timothy Field for his pioneering work on the stochastic differential equation (SDE) theory of sea clutter.

Needless to say, the entire work described in this book would not have been possible without the sustained financial support provided by the Natural Sciences and Engineering Research Council (NSERC) of Canada, for which I am grateful.

I am grateful to George Telecki, Associate Publisher, and Rachel Witmer, Editorial Coordinator, for their full support and help in launching this book. In particular, I would like to express my deep gratitude to Danielle Lacourciere, Senior Production Editor II, STM Book Production, John Wiley, for her hard work and dedication in the actual production of the book.

Last, but by no means least, I am indebted to Lola Brooks, my Technical Coordinator, for working with me for two decades and for taking care of the typing and preparation of the manuscript for the book.

Simon Haykin
Ancaster, Ontario, Canada
July, 2006

Contributors List

Anastasios Drosopoulos
Prof. Electrical Engineering
Patras Institute of Technology (TEI Patras)
M. Alexandrou 1
26334 Patras, Greece

Rembrandt Bakker
Oppermoeren 12, 4824KH, Breda
The Netherlands

Brian Currie
6 Rankin Bridge Rd. (RR3)
Wiarton, Ontario
N0H 2T0

Fulvio GINI
University of Pisa
Department of "Ingegneria dell'Informazione"
via G.Caruso 14
56122 PISA, Italy

Maria V. Sabrina Greco
Dept. of "Ingegneria dell'Informazione"
University of Pisa
Via G.Caruso
56122 Pisa - Italy

Simon Haykin
McMaster University
Adaptive Systems Laboratory, CRL-103
1280 Main Street West
Hamilton, ON Canada L8S 4K1

Dr. David J. Thomson
Queens University
Dept. of Mathematics and Statistics
Kingston, ON K7L 3N6

Chapter 1

Introduction

Simon Haykin

Radar is an active sensor that operates by transmitting an electromagnetic signal and then processing the *radar returns* (i.e., echoes from the many and diverse objects that constitute the surrounding environment). The radar application of interest has a direct bearing on two related issues:

- Specification of the transmitted signal
- Processing of the radar returns

There is no single framework for either one of these two issues; rather, the radar application dictates the way in which the framework is implemented.

In this book, we focus on two types of radar, as summarized here:

1. *Surveillance radar*, the purpose of which may be that of target detection.[1] In *target detection*, the requirement is to *detect* the presence of a moving target (e.g., aircraft or fishing boat) in the presence of unwanted signals in a reliable manner. The unwanted signals consist of *clutter* (i.e., radar backscatter from objects other than the target that lie in the path of the transmitted radar signal), *interference* (i.e., electromagnetic signals produced by other nearby transmitters that could be operating in the same band as the radar transmitter itself), and the ubiquitous *noise* produced by electronic devices at the front end of the receiver.

[1] A related application of surveillance radar is *target classification*, where the requirement is to reliably classify the various objects that constitute the radar environment. For example, in an air traffic control environment, we may be required to distinguish between different objects: aircraft, weather, migrating flocks of birds, and ground. In such an application, the type of radar clutter assumes the role of a target of interest. This application is described in the paper: S. Haykin, W. Stehwien, C. Deng, P. Weber and R. Mann (1991), "Classificaiton of Radar Clutter in an Air Traffic Control Environment", Proc. IEEE, vol. 79, No. 6, pp. 742–772.

Adaptive Radar Signal Processing. Edited by Simon Haykin

2. *Low-angle tracking radar*, the purpose of which, for example, may be that of *tracking* a sea-skimming missile. In such an application, the task of tracking the missile is complicated by the presence of *multipath* caused by reflections from the sea/ocean surface. The multipath problem becomes particularly severe when the missile lies in close proximity to the sea/ocean surface, in which case the radar designer is confronted with having to design a signal-processing algorithm that can reliably distinguish between the missile and its *image* formed below the sea/ocean surface. In a loose sense, the presence of multipath plays a role similar to that of clutter in target detection.

Irrespective of whether the issue of interest is that of target detection, classification, or tracking, solution of the problem is complicated by the *nonstationary character* of the received radar signal. The causes of nonstationarity include motion of the target(s) and variations in environmental conditions. To deal with this complication, we resort to the use of *adaptive radar signal processing*, which is the very title of the book.

EXPERIMENTAL RADAR FACILITIES

Much of the experimental results presented in chapters 2 through 6 are based on real-life radar data collected in a marine environment using two quality-instrument radar facilities. These two facilities are briefly described in what follows. The carefully ground-truthed data collected with these facilities made it possible to test new radar signal-processing algorithms, develop new experimental techniques, and discover new models, all in the context of a marine environment. Moreover, the data were shared with researchers all over the world, which has made construction of the facilities all the more satisfying.

MARS[2]

The primary motivation behind this experimental facility was to collect multipath data representative of a low-elevation target located over water, which would allow the evaluation of *high-resolution angle-of-arrival estimation procedures* [1, 2]. The goal was to design a large array (consisting of 32 elements), which would provide great accuracy and operate over a wide variety of surface roughness (encompassing both specular and diffuse kinds of multipath). In particular, the system would be sufficient for the evaluation of high-accuracy/high-resolution estimation algorithms.

The operating frequency for MARS was 9.81 GHz, providing a free-space wavelength of approximately 3.05 cm. Figure 1.1 shows a block diagram of the transmitter, which consists basically of the following components:

- free-running 5 MHz double-oven crystal oscillator, which is phase-locked up to 9.81 GHz;

[2] "MARS" is abbreviation for "Multi-parameter Adaptive Radar System".

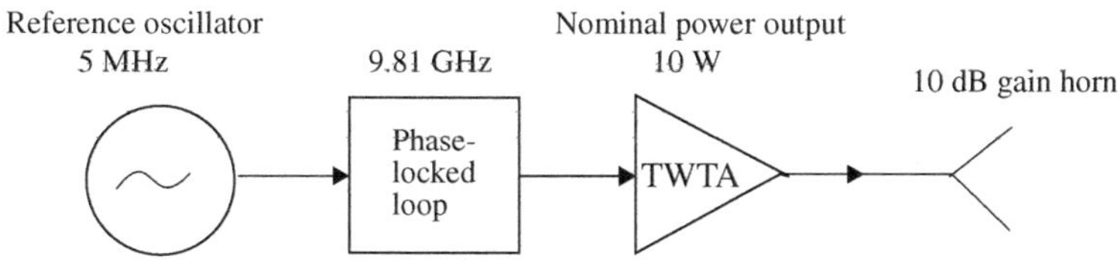

Figure 1.1 Block diagram of the transmitter. (TWTA is abbreviation for travelling-wave tube amplifier)

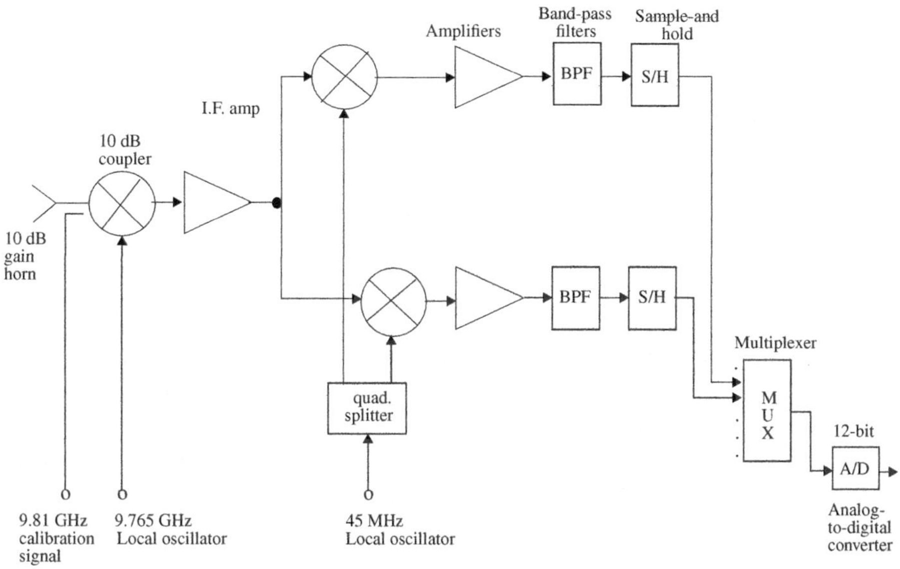

Figure 1.2 Block diagram of one of 32 channels in the receiver

- travelling-wave tube (TWA) amplifier, providing 10 watts of output power; and
- transmitting antenna, consisting of a 10-dB gain horn.

The 5 MHz crystal oscillator provides very low phase-noise; after a 24-hour burn-in time, the long-term drift of the oscillator was quoted to be less than 3 parts in 10^{10} per day.

The receiver consists of a 32-element uniformly spaced aperture; Fig. 1.2 shows a block diagram of the receiver. The front end of each channel of the receiver consists of a 10-dB gain horn, followed by a 10-dB directional coupler. A test signal (used for calibration) could be injected into the system through this coupler when the transmitter is shut down. The received signal/test signal is mixed down to approximately 45 MHz and then amplified. Next, the path is split and mixed down to "in-phase" and "quadrature baseband signals" having frequencies of 15.625 Hz. After further amplification and low-pass filtering (with cutoff frequency at 31.25 Hz), the resulting signal is sampled at 125 Hz, that is, 8 samples per cycle.

The low-frequency signals were all digitally generated, synchronous with a computer system clock. The baseband frequency, filter bandwidths and sampling rates could all be varied under computer control, if experimental conditions required it.

Prior to data collection, with the transmitter on, the 5 MHz oscillator at the receiver was fine-tuned such that the receiver was operating within 0.1 Hz of the transmitted signal at X-band. In effect, the receiver could be viewed as essentially "coherent" for the duration of data collection, usually less than 10 seconds. For long-term data collection, provision was made for continuous adjustment of the 5 MHz oscillator. When the test signal was applied instead of the received signal, the system was fully coherent. In particular, coherence of the system would allow for extremely fine Doppler measurements due to motion of the water surface.

The linear array at the front end of the receiver was oriented for vertical polarization. In more specific terms, the structure of the array was machined such that the spacing between the 32 horns of the array is 5.715 ± 0.010 cm. A similar tolerance was attained for the remaining two horizontal dimensions of the array structure. The electrical phase error with respect to neighboring elements (horns) was less than 1°. The unambiguous field of view was approximately ± 15.5°. In terms of normalized parameters where the spacing between elements is considered equal to one unit, the span of wavenumber estimation achievable by the array is $\pm\pi$, with π corresponding to a physical elevation angle of 15.5°. With the physical aperture of the 32-element array structure being 1.77 m, the beamwidth in physical terms is approximately 1°.

In building the receiver, precautions were taken to ensure that the 32 channels of the receiver will respond to environmental changes in similar ways. Moreover, immediately before and after a set of real-life radar data is collected, an electronic calibration of the system was performed. Typically, the total time from initial electronic calibration, followed by data collection and one final calibration, was less than 30 minutes.

The low-angle tracking experiments were performed on the mouth of Dorcas Bay, which opens into the eastern end of Lake Huron, Ontario. As illustrated in Fig. 1.3, the transmitter was located at a distance of 4.75 km from the receiver, both being within 10 m from the water's edge; occasionally, during storms, the transmitter and receiver would be within the Bay itself when the water level rose.

IPIX[3] Radar

The IPIX radar is a transportable, digitally controlled, coherent dual-polarized X-band radar, designed to be of instrument-quality for research use [3, 4]. The radar

[3] Development of the IPIX radar began in 1984, and a prototype version of the system was tested in the summer of 1986 at Cape Bonnavista, Newfoundland. Originally, the IPIX radar was shorthand for "Ice multi-Parameter Imaging X-band" radar, so called as the radar was designed for the detection of growlers (i.e., small pieces of ice broken off an iceberg). After major upgrades to the radar, which were carried out between 1993 and 1998, the high-resolution data collected by the IPIX radar became a benchmark for testing intelligent detection algorithms. Accordingly, the meaning of the IPIX radar was changed into "Intelligent PIxel processing X-band" radar, where the term "pixel" refers to a picture element.

It is also noteworthy that the cover of the book pictures the antenna of the IPIX radar.

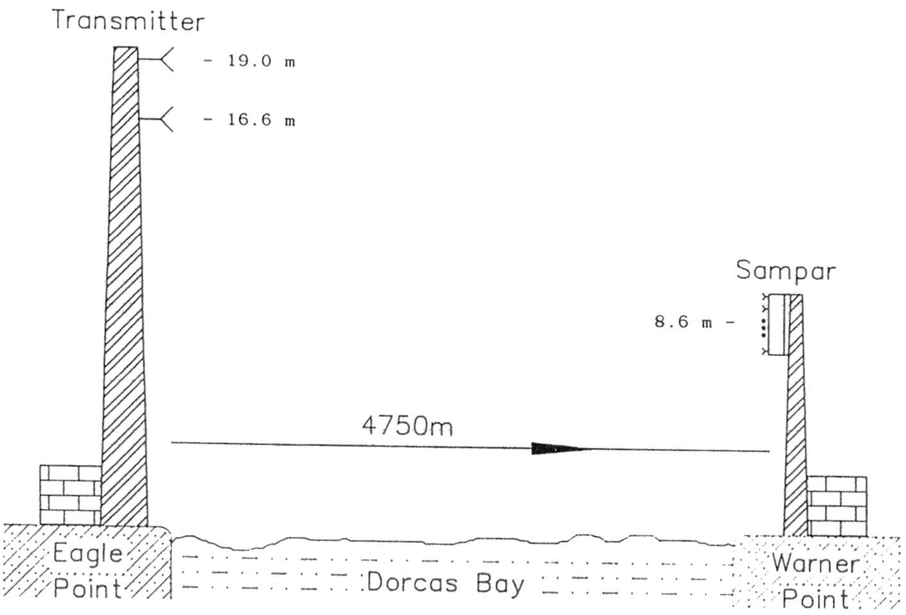

Figure 1.3 Experimental site description (not to scale); sampar is abbreviation for "sampled aperture."

was built as part of an extensive research program aimed at the development of improved technologies/algorithms for the detection and identification of small targets in an ocean environment. The foundation for the research program was perceived to develop a thorough understanding of the nature of *sea clutter* and the corresponding behavior of targets of interest under varying sea conditions. To this end, the IPIX radar was built to collect a database of sea clutter and target radar returns so as to characterize the ocean environment and target behavior with respect to the radar parameters. The major data collections were performed at two cities: Cape Bonavista, Newfoundland and Dartmouth, Nova Scotia, both on the Atlantic Coast Table 1.1 summarizes the radar parameters.

ORGANIZATION OF THE BOOK

The book is organized in two parts.

Part I, consisting of two chapters, deals with *radar spectral analysis.* As the name implies, the primary focus of attention here is estimation of *spectrum* of the received signal. With the objectives of the two chapters being different, the tools used to perform the spectral analysis are correspondingly different.

Chapter 2 addresses the low-angle tracking radar problem. With estimation of the target's *angle of arrival (AOA)* in the presence of multipath as the issue of interest, we use a spectrum estimation procedure known as the *multi-taper method* or *multiple-window method.* This method was originally formulated in the time domain. On the other hand, the low-angle tracking radar problem is of a spatial

Table 1.1 Major Features of the IPIX Radar System

Transmitter
- 8-kW peak power TWT
- H or V polarization, switchable pulse-to-pulse
- Frequency fixed (9.39 GHz) or agile over 8.9–9.4 GHz
- Pulse width 20–200 ns (20-ns steps), 200–5000 ns (200-ns steps)
- Pulse repetition frequency up to 20 kHz, limited by duty cycle (2%) or polarization switch (4 kHz)
- Pulse repetition interval, configurable on a per-pulse basis

Receiver
- Fully coherent reception
- Two linear receivers; H or V on each receiver (usually one H, one V for dual-polarized reception)
- Instantaneous dynamic range >50 dB
- 8-bit, or 10-bit with hardware integration, sampling
- 4 A/Ds: I and Q for each of two receivers
- Range sampling rate up to 50 MHz
- Full-bandwidth digitized data saved to disk, archived onto CD

Antenna
- 2.4-m-diameter parabolic dish
- Pencil beam, beamwidth 0.9°
- 44-dB gain
- Sidelobes < –30 dB
- Cross-polarization isolation
- Computer controlled positioner
- –3° to +90° in elevation
- Rotation through 360° in azimuth, 0–10 rpm

General
- Radar system configuration and operation completely under computer control
- User operates radar within an IDL environment

kind, hence the need for reformulating the method as one of *wavenumber spectrum estimation*. Most important, the multi-taper method accounts for the *specular* as well as *diffuse* kinds of multipath, which are integral parts of a physical low-angle tracking radar environment. Most important, this method deals with the problem in a composite and highly elegant manner.

Chapter 3 also uses the multi-taper method, but with an important extension. Specifically, the power spectrum is now estimated as a *function of both time and frequency*. Moreover, the application of interest is the characterization of radar returns produced in a marine environment, with the objective of discriminating between target returns and sea clutter (i.e., radar backscatter from the sea/ocean surface).

Both Chapters 2 and 3 not only present mathematical details of the algorithms used to perform the spectral analysis but also include experimental results based on real-life radar data collected with the MARS and IPIX systems.

Part II of the book, consisting of Chapters 4 through 6, deals with *dynamic models* of radar returns produced in a marine environment.

Chapter 4 focuses on modeling the underlying dynamics responsible for the generation of sea clutter. Three specific approaches are discussed in this chapter:

- *Chaos* as a possible mechanism for describing sea clutter; here we look to chaos theory applied to sea clutter data to test the applicability of this theory.
- *Hybrid amplitude modulation-frequency modulation*, the use of which is motivated by the underlying physics of sea clutter published in the literature.
- *Autoregressive (AR) model*, the parameterization of which follows well-established statistical estimation theory.

Chapter 5 expands on the ideas described on modulation theory in Chapter 4 and thereby further refines this physical basis for the statistical characterization of sea clutter dynamics by accounting for nonstationarity.

Chapter 6 completes the discussion on dynamics of radar returns in a marine environment by formulating a *Bayesian framework for detection-through-tracking* of a target (moving on the sea surface) in the presence of sea clutter. Unlike classical detection theory based on hard decisions, the information content of radar returns is preserved through the use of soft decisions.

As with Part I of the book, the adaptive signal-processing theory presented in all three chapters of Part II is supported experimentally using real-life radar data collected using the IPIX radar under different environmental conditions.

REFERENCES

1. E. J. Vertatschitsch (1987). *Linear Array for Direction of Arrival Estimation.* Ph.D. Thesis, McMaster University, Hamilton, Ontario.
2. A. Drosopoulos (1992). *Investigation of Diffuse Multipath at Low Grazing Angles.* Ph.D. Thesis, McMaster, University, Hamilton, Ontario.
3. S. Haykin, C. Krasnor, T. J. Nohara, B. W. Currie, and D. Hamburger (1991). A coherent dual-polarized radar for studying the ocean environment. *IEEE Trans. Geoscience and Remote Sensing*, **29**(1), 189–191.
4. S. Haykin, B. W. Currie, and V. Kezys (1994). Surface-based Radar: Coherent. In: *Remote Sensing of Sea Ice and Icebergs*, S. Haykin, E. O. Lewis, R. K. Raney, and J. R. Rossiter (editors), Wiley, 443–504.

Part I

Radar Spectral Analysis

Chapter 2

Angle-of-Arrival Estimation in the Presence of Multipath†

Anastasios Drosopoulos and Simon Haykin

2.1 INTRODUCTION

This chapter deals with the angle-of-arrival estimation problem, which may be viewed as a problem in *wavenumber spectrum estimation* (i.e., spectrum estimation in the spatial domain). Consider, for example, an ordinary *low-angle tracking radar* engaged with a sea-skimming missile. At low grazing angles, the proximity of the target to the sea surface gives rise to the well-known *multipath phenomenon*, the description of which depends on the condition of the sea surface, as summarized here:

- In the idealized case of a perfectly *smooth surface*, the multipath model consists of two components. One component, lying above the surface and referred to as the *direct component*, originates from the target itself. The second component, lying below the surface and referred to as the *specular component*, gives rise to an *image* target. The received signal from this image is related to the actual target signal by the *reflection* or *Fresnel coefficients*. This idealized model is referred to as *specular multipath*.
- A more accurate model of the multipath phenomenon accounts for the unavoidable surface *roughness*, which has the effect of modifying the specular component. Furthermore, nonspecular components, constituting *diffuse multipath*, are introduced into the composition of the overall received signal.

† The material presented herein is based on the following chapter contribution: A. Drosopoulos and S. Haykin (1992) "Adaptive radar parameter estimation with Thomson's Multiple-Window Method," in S. Haykin and A. Steinhardt (eds.), *Adaptive Radar Detection and Estimation*, Wiley, New York, pp. 381–461.

Adaptive Radar Signal Processing. Edited by Simon Haykin

In any event, since the specular component and (to a lesser degree) the diffuse component are correlated with the direct component (representing the desired target signal), *fading* can occur. Consequently, in an extreme case, it is possible for the desired target signal to be *canceled* because of phase opposition introduced into the picture by multipath. Indeed, due to practical limitations imposed on the physical size of the radar antenna's aperture, the direct and specular components may enter the antenna's main lobe, which, in turn, would make the task of resolving the two components extremely hard. Moreover, the presence of diffuse multipath may further complicate the angle-of-arrival estimation problem.

In this chapter, we approach the solution to this difficult estimation problem by using the *multi-taper method*, which was first described by David Thomson in 1982. In his original paper, the spectrum estimation procedure (formulated in the time domain) was referred to as the *method of multiple windows*. The multi-taper method, rooted in classical spectrum estimation theory, is not only mathematically elegant, but has also impacted many physical disciplines outside of signal processing.

The chapter also includes experimental results, using computer simulations and real-life multipath data. Specifically, a 32-element uniformly sampled aperture *multiparameter adaptive radar system*, dubbed MARS,[1] was used for data collection on a site located at Lake Huron, Ontario. To simplify the multipath data collection at low grazing-angle conditions over the lake surface, the system was designed to operate in a *bistatic* mode, with a separate transmitter posing as the "target" of interest and the sampled-aperture antenna (Sampar) operating as the "receiver." In this way, the received signal consists of the desired target (i.e., direct) component and multipath (including specular as well as diffuse components). Except for the ubiquitous receiver noise, there is no other clutter component.

Modern signal-processing techniques that promise *increased spatial resolution* are examined in this chapter. However, this increased resolution can only be achieved by having accurate models of the phenomena involved in order to be able to estimate

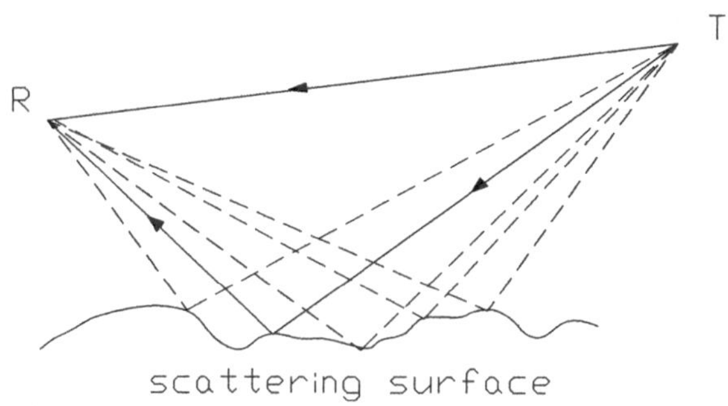

Figure 2.1 Illustrating the variety of paths (multipath) that a signal from a target (T) can reach the radar receiver (R) over a water surface.

[1] MARS is described in Chapter 1.

the desired parameters (directions of arrival) from the data. Model simplicity usually assumes the existence of specular multipath only, ignoring diffuse multipath. Theoretical models of vector electromagnetic wave rough surface scattering can become quite complex at low-grazing angles when shadowing and diffraction effects are significant. In fact, a full solution has yet to be developed.

In the study reported herein, a relatively new, nonparametric technique is developed that estimates in an optimum and data-adaptive manner the spatial/temporal (wavenumber/frequency) characteristics of the received signal *without a priori model assumptions*. Our goal is first to describe the method in sufficient detail as to make it informative and useful to implement in any situation where accurate experimental spectra are desired. Second, we describe our results of applying this method to angle-of-arrival estimation at low-grazing angles with diffuse multipath taken into account. Finally, we compare some particular theoretical spectra with measured results.

2.2 THE LOW-ANGLE TRACKING RADAR PROBLEM

Data-adaptive parameter estimation refers to the estimation of one or more of the radar parameters in an adaptive manner. In this chapter we focus on the *angle-of-arrival (AOA) estimation* problem. The instrumentation most suited for this purpose is a *sampled-aperture antenna* where concurrent data samples are taken at different points in space and subsequently combined in a suitable manner (beamforming) to estimate the AOA of an incoming signal. Data-adaptivity comes in the signal processing involved with the beamforming process.

The simplest form of beamforming is to use all the data samples to construct the wavenumber spectrum (analogous to the frequency spectrum of a time series), which can be one- or two-dimensional, depending on whether the sampled aperture is one- or two-dimensional. In essence the sampled aperture can resolve incoming signals in a direction perpendicular to the aperture axis (a vertical aperture resolves elevation angles and a horizontal aperture resolves azimuth angles). The resolution limit is on the order of a beamwidth (which, for an M-element linear array, is defined as $2\pi/M$). The more sensors the aperture has, the better the resolution capability will be. Superresolution techniques can, in principle, achieve better performance than traditional techniques based on Fourier transforms at the cost of more intense signal processing.

At low grazing angles, the situation becomes particularly difficult because, on the one hand, the direct and specular components are both at sub-beamwidth separations; and also the specular component, being coherent with the direct, can approach phase opposition with it, thereby causing signal cancellation. This is particularly severe for the common monopulse radar, which is normally incapable of performing well in a multipath environment, with the radar losing track of a low-altitude flying target.

The most common way to deal with this problem is to perform beamforming of the received signal in order to suppress signals coming from directions below the horizon, so as to ensure that the main lobe of the receiver antenna is always pointing above the horizon and to suppress sidelobe signals residing in the sidelobes. This is where adaptivity becomes useful, provided that the radar is designed to adapt its beamforming technique in accordance with the received data.

Practical issues (such as cost and physical limitations) put constraints on the number of elements in a sampled aperture and the separation between them. *Thomson's multi-taper method* (MTM), described in this chapter, is shown to perform in a robust enough manner for the difficult case of correlated signals and to extract the desired signal information in an optimum way.

2.3 SPECTRUM ESTIMATION BACKGROUND

Consider the sequence $\{x(t_i)\}_{i=1}^{N}$ composed of sequential samples from a single realization of a complex-valued, weakly stationary (i.e., wide-sense), continuous-time, one-dimensional[2] random process $X(t)$. We further assume that the process is zero-mean with autocorrelation function $r_x(\tau)$; that is, using $\mathbf{E}$ to denote the statistical expectation operator (a notation followed throughout the book), we have

$$\mathbf{E}\{X(t)\} = 0$$

and

$$\mathbf{E}\{X^*(t)X(t+\tau)\} = r_x(\tau) \tag{2.1}$$

If the mean $\bar{X}(t) = \mathbf{E}\{X(t)\} \neq 0$, then the above relations are satisfied by the difference $X(t) - \bar{X}(t)$.

The definition of the power spectrum in terms of the autocorrelation function is given by the *Wiener–Khintchine relations*

$$r_x(\tau) = \int_{-\infty}^{+\infty} S(f)e^{j2\pi f\tau}df \quad \text{and} \quad S(f) = \int_{-\infty}^{+\infty} r_x(\tau)e^{-j2\pi f\tau}d\tau \tag{2.2}$$

where $S(f)$ is the *power spectral density* (PSD) or simply spectrum of the process $X(t)$. $S(f)\,df$ represents the average (over all realizations) contribution to the total power (or process variance) from all possible components of $X(t)$ with frequencies lying between f and $f + df$.

The power interpretation of the spectrum is more evident from an alternative definition expressed in terms of the random process itself [27]:

$$S(f) = \lim_{T\to\infty} \mathbf{E}\left[\frac{1}{T}\left|\int_{-T/2}^{T/2} X(t)e^{j2\pi ft}dt\right|^2\right] \tag{2.3}$$

Typical nonparametric spectrum estimators for finite data may be considered approximations to either (2.2) (Blackman and Tukey method [3]) or (2.3) (modified

[2] Generalization of the treatment to more dimensions is done by simply allowing time t to become a d-dimensional vector $\mathbf{t}$, where d is the dimension of the process. See references 5 and 6 for a concise explanation. In the following, we consider one-dimensional processes only, since they adequately describe our experimental data.

periodogram). The Blackman and Tukey spectrum estimate, for a data sample of size N, is given by the formula

$$\hat{S}(f) = \sum_{m=1-N}^{N-1} \hat{r}(\Delta m) d(m) e^{-j2\pi f \Delta m}$$

where the autocorrelation sequence is estimated as

$$\hat{r}(\Delta m) = \frac{1}{N-m} \sum_{n=1}^{N-m} x[\Delta(n+m)] x^*(\Delta n)$$

where $0 \le m \le N - 1$, $\hat{r}(\Delta m) = r^*(-\Delta m)$ for $m < 0$, and Δ is the sampling period. The weight sequence $\{d(n)\}$ has real positive elements satisfying $d(m) = d(-m)$ to ensure that the spectral estimate is real, and $d(0) = 1$ to ensure that it is unbiased when the true spectrum is flat across $\mathcal{B}$, where $\mathcal{B} = (f\text{: } |f| < 1/2\Delta\}$.

The modified periodogram spectrum estimate is given by

$$\hat{S}(f) = \left| \sum_{n=1}^{N} x(\Delta n) c(n) e^{-j2\pi f \Delta n} \right|^2$$

where $f \in \mathcal{B}$ and the weight sequence $\{c(n)\}$ typically has real, positive elements satisfying $\Sigma_{n=1}^{N} c^2(n) = \Delta$ to ensure that the spectral estimator is unbiased when the true spectral density is flat across $\mathcal{B}$. To approximate the ensemble—averaging (statistical) operator **E**, the data record is usually segmented and individual results are averaged to reduce estimator variance.

Usually, the underlying distribution of $X(t)$ is assumed to be Gaussian, so that the second-order statistics suffice for a complete description of the process. Otherwise, a higher order statistics hypothesis should be made (see reference 37 for bispectrum estimation using multiple windows). The ergodicity property, which holds for a zero-mean Gaussian process *with no line components*, is also frequently invoked, so that ensemble averages can be replaced by time averages.

However, interest in classical spectrum estimation was renewed [25] only after the publication of Thomson's classic 1982 paper [36], where the power of MTM is demonstrated. Basically, Thomson has proved that a more fruitful approach to a spectrum estimator is through the spectral representation of $X(t)$ itself (Cramér representation). Ishimaru [12] gives a particularly lucid explanation of how this representation is defined. Following Ishimaru's arguments, consider a stationary complex random function $X(t)$ that satisfies (2.1). In attempting to develop a spectral representation for the random function $X(t)$, it is tempting to write down the Fourier transform

$$X(t) = \int_{-\infty}^{\infty} X(f) e^{j2\pi f t} df$$

However, the stationarity assumption is then violated, since Dirichlet's condition requires that $X(t)$ be absolutely integrable—that is, that $\int_{-\infty}^{\infty} |X(t)|\, dt$ be finite. To avoid this difficulty, the random function is represented by a *stochastic Fourier–Stieltjes integral,* as shown by

$$X(t) = \int_{-\infty}^{\infty} e^{j2\pi ft}\, dZ(f)$$

where $dZ(f)$ is called the *random amplitude* or *increment process.* To determine the properties of $dZ(f)$, we examine (2.1). First, we require that

$$\mathbf{E}\{dZ(f)\} = 0$$

and second that the covariance[3] function

$$\mathbf{E}\{X(t_1)X^*(t_2)\} = \int_{-\infty}^{\infty}\int_{-\infty}^{\infty} e^{j2\pi f_1 t_1 - j2\pi f_2 t_2}\,\mathbf{E}\{dZ(f_1)\,dZ^*(f_2)\}$$

be a function of the time difference $t_1 - t_2$ only. This second condition requires that we write

$$\mathbf{E}\{dZ(f_1)\,dZ^*(f_2)\} = S(f_1)\delta(f_1 - f_2)\,df_1 df_2 \tag{2.4}$$

where $S(f)$ is the power spectral density of the process, representing the amount of power density at different frequencies. It is identical to the definition given previously, as can be seen by substituting the covariance (2.4) in the Wiener–Khintchine relations (2.3). Note also that at different frequencies, $\mathbf{E}\{dZ(f_1)dZ^*(f_2)\}$ is zero; that is, the increments $dZ(f)$ are orthogonal (the energy at different frequencies is uncorrelated).

For the discrete-time case, the Wiener–Khintchine relations become

$$r_x(n) = \int_{-1/2}^{1/2} S(f)e^{j2\pi fn}\,df \quad \text{and} \quad S(f) = \sum_{n=-\infty}^{\infty} r_x(n)^{-e^{j2\pi fn}} \tag{2.5}$$

where $n = 0, \pm 1, \ldots$ and the time between samples is taken to be 1, so that the frequency f is confined in the principal domain $(-\frac{1}{2}, \frac{1}{2}]$. Similarly, the discrete-time spectral representation of the time series $\{x(n)\}$ is given by

$$x(n) = \int_{-1/2}^{1/2} e^{j2\pi fn}\,dZ(f) \tag{2.6}$$

The spectral representation concept is quite basic; in essence, it says that any stationary time series can be interpreted as the limit of a finite sum of sinusoids $A_i\cos(2\pi f_i t + \Phi_i)$ over all frequencies $f = f_i$. The amplitudes $A_i = A(f_i)$ and phases $\Phi_i = \Phi(f_i)$ are uncorrelated random variables with $S(f) \approx \mathbf{E}\{A^2(f_i)\}$ for $f_i \approx f \neq 0$ and can be related to $dZ(f)$ in a simple way (see reference 19). This point of view leads us to write

$$\mathbf{E}\{dZ(f)\} = 0 \quad \text{and} \quad S(f)\,df = \mathbf{E}\left\{|dZ(f)|^2\right\} \tag{2.7}$$

as the proper definition of the power spectrum.

When a number of line components is present, the above relations are easily generalized to include them. The first moment becomes

$$\mathbf{E}\{dZ(f)\} = \sum_i \mu_i \delta(f - f_i)\,df \tag{2.8}$$

[3] The terms autocorrelation and covariance are equivalent for zero-mean processes.

where f_i are the frequencies of the periodic or line components, and μ_i are their amplitudes. The *continuous part of the spectrum*, or second moment, becomes

$$S(f)df = \mathbf{E}\left\{|dZ(f) - \mathbf{E}\{dZ(f)\}|^2\right\} \tag{2.9}$$

First moments are associated with the study of periodic phenomena (harmonic analysis). Typically, few such lines will exist in a process, each being described by its amplitude, frequency, and phase.[4] Such parameters may be estimated using techniques based on maximum-likelihood. In classic methods of spectrum estimation —nonparametric, based on the periodogram—the resolution limit (also referred to as the *Rayleigh resolution limit*) is $1/T$, where T is the total observation time. Superresolution—that is, the ability to discriminate frequencies spaced closer than the Rayleigh resolution—is possible, depending on the SNR, which is defined as the ratio of power in the first moment to power in the second moment as a function of frequency.

Second moments, on the other hand, are stochastic in character. In contrast to the line spectrum of the first moment, the second moment spectrum is typically continuous and often smooth. In this case, the issue of interest is to estimate a function of frequency, not just a handful of parameters, and therefore maximum-likelihood parameter estimation is not applicable here. Concerning resolution, it is impossible now to resolve details separated by less than twice the Rayleigh limit. Typically, resolution is between $2/T$ and $50/T$, much poorer than Rayleigh.

Thomson [39] states that "confusing the distinction between the two moment properties will result in absurdities like smoothing line spectra or applying superresolution criteria to noise-like processes." Finally, a point that must be kept in mind is the fact that classically, spectra are defined only for stationary processes. For nonstationary processes the usual assumption that allows us to work with them is that of local stationarity.

2.3.1 The Fundamental Equation of Spectrum Estimation

We assume that we have a finite data set of N contiguous samples, $x(0), x(1), \ldots,$ $x(N-1)$, which are observations from a stationary, complex, ergodic, zero-mean, Gaussian process. The problem of spectrum analysis is that of estimating the statistical properties of $dZ(f)$ from the finite time series

$$\{x(n)\}_{n=0}^{N-1}$$

Taking the Fourier transform of the data, we obtain

$$\tilde{x}(f) = \sum_{n=0}^{N-1} x(n)e^{-j2\pi fn} \tag{2.10}$$

[4] For an excellent treatment of decaying sinusoids with the multiple-window method, see references 29 and 30.

and substituting the Cramér representation for $x(n)$, we arrive at the fundamental equation of spectrum estimation

$$\tilde{x}(f) = \int_{-1/2}^{1/2} \mathbf{D}_N(f - \nu)\, dZ(\nu) \tag{2.11}$$

where the kernel is given by

$$\mathbf{D}_N(f) = \sum_{n=0}^{N-1} e^{-j2\pi fn} = \exp[-j\pi f(N-1)]\left(\frac{\sin N\pi f}{\sin \pi f}\right) \tag{2.12}$$

We may now interpret the fundamental equation as a convolution that describes the window leakage or smearing that is a consequence of the finite sample size. Clearly, there is no obvious reason to expect the statistics of $\tilde{x}(f)$ to resemble those of $dZ(f)$.

The fundamental equation may be viewed as a *linear Fredholm integral equation of the first kind for* $dZ(f)$. Since this is the frequency-domain expression of the projection from an infinite stationary process generated by the random orthogonal measure $dZ(f)$ onto the finite observation sample, it does not have an inverse. This makes it impossible to find exact or unique solutions. Instead, our goal becomes one of searching for approximate solutions whose statistical properties are, in some sense, close to those of $dZ(f)$.

The above observation is another way of saying that the problem of spectrum estimation from finite data is an *ill-posed inverse problem*. Mullis and Scharf [25] define both the time-limiting operation (windowing, finite data) and isolation in frequency (power in a finite spectral window) as projection operators on the data, $\mathbf{P}_T$ and $\mathbf{P}_F$, respectively. These operators do not commute, that is, $\mathbf{P}_T\mathbf{P}_F \neq \mathbf{P}_F\mathbf{P}_T$. If they did, then their product would also be a projection and it would then be possible to isolate a signal component in both time and frequency. However, under certain conditions, $\mathbf{P}_T\mathbf{P}_F \approx \mathbf{P}_F\mathbf{P}_T$ and the product operator is close to a projection having rank NW, denoting the *time–bandwidth product*. It turns out that Thomson's MTM is equivalent to a projection of the data onto a subspace where the signal power in a narrow spectral band is maximized; that is, the conditions required for the above-mentioned operators to approximately commute are found.

2.4 THOMSON'S MULTI-TAPER METHOD

Generally, in a radar environment the received signal is composed of a desired direct signal(s) coming from the target(s) of interest, their multipath reflections, and/or clutter plus receiver noise. In detection and tracking, the need exists to accurately estimate the angle(s) of arrival (AOA) of the desired signal(s). Data from a sampled-aperture antenna lead to the estimation of the wavenumber spectrum as well. *Multipath* is commonly divided into two kinds: *specular* and *diffuse*. The first, being essentially a plane wave, appears as an additional spectral line, while the latter, being stochastic, has a broader continuous shape. The need exists therefore to somehow estimate this mixed spectrum in the best way possible, from a finite set of data samples. Ideally, the method used for this purpose should be nonparametric in nature (this refers to the continuous spectrum background) so as not to be influenced by any *a priori* assumption about the signal structure.

Thomson's multi-taper method (MTM) as expanded in references 36, 38, and 39 is our method of choice to solve this estimation problem.[5] In particular, MTM offers the following attractive features:

- MTM is nonparametric.
- It provides a unified approach on spectrum estimation.
- It is optimally suited for finite-data samples.
- It can be generalized to irregularly sampled multidimensional processes [5, 6].
- It is consistent and efficient.
- It has good bias control and stability.
- It also provides an analysis of variance test for line components.

The solution of the fundamental equation (2.11) is found in terms of the eigenfunction expansion of the kernel, which is recognized as the *Dirichlet kernel with known eigenfunctions*, namely, the *prolate spheroidal wavefunctions*; these functions are fundamental to the study of time and frequency-limited systems. Exploiting this finite-bandwidth property of the process to be estimated, a search for solutions to (2.11) is carried out in some local interval about f, say $(f - W, f + W)$, using the prolate spheroidal wavefunctions as a basis.

2.4.1 Prolate Spheroidal Wavefunctions and Sequences

From Slepian,[6] the eigenfunction expansion of the Dirichlet kernel [33] is given by

$$\int_{-W}^{W} \frac{\sin N\pi(f-f')}{\sin \pi(f-f')} U_k(N, W; f')df' = \lambda_k(N, W)U_k(N, W; f) \qquad (2.13)$$

where the $U_k(N,W;f)$, $k = 0, 1, \ldots, N-1$ are the *discrete prolate spheroidal wavefunctions* (DPSWF); and W, lying in the interval $0 < W < \frac{1}{2}$, is the *local bandwidth*, which is normally on the order of $1/N$.

[5] Useful background information on the MTM and implementation examples are also given in references 8, 19, 20, 23, 28, 29, and 35; some recent work reported in references 13 and 26 also offers a point of view that is more familiar to the array signal processing community.

[6] In the most recent literature, the prolate spheroidal wavefunctions and sequences are also referred to as Slepian functions and Slepian sequences, respectively, to honor David Slepian, who first described their properties in signal processing and statistical applications. The term "prolate spheroidal" first came about from the solution of the wave equation in a prolate spheroidal coordinate system. The zeroth-order solution of that differential equation is also the solution to the integral equation (2.13).

Some of the properties of the discrete prolate spheroidal wavefunctions are as follows:

- The functions are ordered by their eigenvalues, as shown by

$$1 > \lambda_0(N, W) > \lambda_1(N, W) > \ldots > \lambda_{N-1}(N, W) > 0$$

The first $K = 2NW$ eigenvalues are very close to 1.

- The functions are doubly orthogonal, which means that

$$\int_{-W}^{W} U_j(N, W; f) U_k(N, W; f)\, df = \lambda_k(N, W)\delta_{j,k}$$

and

$$\int_{-1/2}^{1/2} U_j(N, W; f) U_k(N, W; f)\, df = \delta_{j,k}$$

where

$$\delta_{j,k} = \begin{cases} 1 & \text{for } k = j \\ 0 & \text{otherwise} \end{cases}$$

is the *Kronecker delta function.*

- Their Fourier transforms are the *discrete prolate spheroidal sequences* (DPSS), as shown by

$$\upsilon_n^{(k)}(N, W) = \frac{1}{\varepsilon_k \lambda_k(N, W)} \int_{-W}^{W} U_k(N, W; f) e^{-j2\pi f[n-(N-1)/2]}\, df$$

or

$$\upsilon_n^{(k)}(N, W) = \frac{1}{\varepsilon_k} \int_{-1/2}^{1/2} U_k(N, W; f) e^{-j2\pi f[n-(N-1)/2]}\, df$$

for $n, k = 0, 1, \ldots, N-1$ and

$$\varepsilon_k = \begin{cases} 1 & \text{for } k \text{ even} \\ i & \text{for } k \text{ odd} \end{cases}$$

The DPSWFs can be expressed as

$$U_k(N, W; f) = \varepsilon_k \sum_{n=0}^{N-1} \upsilon_n^{(k)}(N, W) e^{j2\pi f[n-(N-1)/2]}$$

and the DPSSs satisfy the *Toeplitz matrix eigenvalue equation*

$$\sum_{m=0}^{N-1} \left(\frac{\sin 2\pi W(n-m)}{\pi(n-m)} \right) \upsilon_m^{(k)}(N, W) = \lambda_k(N, W) \upsilon_n^{(k)}(N, W)$$

These two equations give a relatively straightforward way of computing the DPSSs and DPSWFs for moderate values of N. In matrix form, the above eigenvalue equation is written as

$$\mathbf{T}(N, W)\mathbf{v}_{(k)}(N, W) = \lambda_k(N, W)\mathbf{v}_{(k)}(N, W)$$

where

$$\mathbf{v}_{(k)} = \left[\upsilon_0^{(k)}, \upsilon_1^{(k)}, \ldots, \upsilon_{N-1}^{(k)}\right]^T$$

and

$$\mathbf{T}(N,W)_{mn} = \begin{cases} \sin[2\pi W(m-n)]/[\pi(m-n)], & m, n = 0, 1, \ldots, (N-1) \text{ and } m \neq n \\ 2W & \text{for } m = n \end{cases}$$

Both Thomson [36] and Slepian [31–34] give asymptotic expressions for the computation of the DPSSs and DPSWFs; it is probably the complexity of these expressions that would initially discourage people from using the prolate basis. If, however, only the eigenvectors are required, Slepian [33] notes that the DPSSs satisfy a *Sturm–Liouville differential equation*, which leads to

$$\mathbf{S}(N,W)\,\mathbf{v}_{(k)}(N,W) = \theta_k(N,W)\,\mathbf{v}_{(k)}(N,W) \tag{2.14}$$

The matrix $\mathbf{S}$ (N,W) is tridiagonal in the sense that

$$\mathbf{S}(N,W)_{ij} = \begin{cases} \frac{1}{2}i(N-1), & j = i-1 \\ \left(\frac{N-1}{2} - i\right)^2 \cos 2\pi W, & j = i \\ \frac{1}{2}(i+1)(N-1-i), & j = i+1 \\ 0, & \text{otherwise} \end{cases}$$

where $i, j = 0, 1, \ldots, N-1$. Even though the eigenvalues θ_k are not equal to λ_k, they are ordered in the same way and the eigenvectors are the same. Tridiagonal systems are easier than Toeplitz to solve, and this offers a practical way of numerically computing the eigenvectors. In actuality, only a small number of eigenvalues and eigenvectors is needed. Given the eigenvectors, the eigenvalues can then be found[7] from

$$\lambda_k(N,W) = [\mathbf{v}_{(k)}(N,W)]^T\, \mathbf{T}(N,W)\,\mathbf{v}_{(k)}(N,W) \tag{2.15}$$

Note that in our case, Slepian's Dirichlet kernel is modulated by a complex exponential factor, which, in turn, leads to the eigenfunction expansion

$$\int_{-W}^{W} \mathbf{D}_n(f-\nu)V_k(\nu)d\nu = \lambda_k V_k(f) \tag{2.16}$$

where, for notational simplicity, the dependence on N and W has been suppressed. The connection with Slepian's original exposition [39] is established by writing

$$V_k(f) = (1/\varepsilon_k)e^{-j\pi f(N-1)}U_k(-f)$$

[7] Thomson [38] uses the routines BISECT and TINVIT to evaluate the Slepian sequences, and

$$\lambda_k(N,W) = \int_{-W}^{W} |V_k(f)|^2\, df \Big/ \int_{-1/2}^{1/2} |V_k(f)|^2\, df$$

for the eigenvalues. Inclusion of N in the argument arises due to dependence of $V_k(f)$ on N.

so that the Fourier transform of Slepian's DPSSs yields

$$V_k(f) = \sum_{n=0}^{N-1} \upsilon_n^{(k)}(N, W) e^{-j2\pi fn} \tag{2.17}$$

A step-by-step procedure involved in computing the data $\upsilon_n^{(k)}$'s and spectral $V_k(f)$'s windows (Slepian sequences and functions) is summarized as follows:

1. Obtain the first N-point data sample; this specifies N.
2. Select a time–bandwidth product NW; this specifies the analysis window W.
3. Use (2.14) and (2.15) to compute the λ_k's and $\upsilon_n^{(k)}$'s; actually, only the first $K = 2NW$ terms with the largest eigenvalues are needed (Thomson [38] suggests the use of $K = 2NW - 1$ to $K = 2NW - 3$ to minimize higher order window leakage).
4. Finally, use (2.17) with the fast Fourier transform (FFT) algorithm (preferably zero-padded) to compute the corresponding $V_k(f)$'s,

Figures 2.2 and 2.3 show an example of data and spectral windows (Slepian sequences and functions) that are used in the test dataset (discussed in the next section). Note that the importance of using these windows is the fact that they are

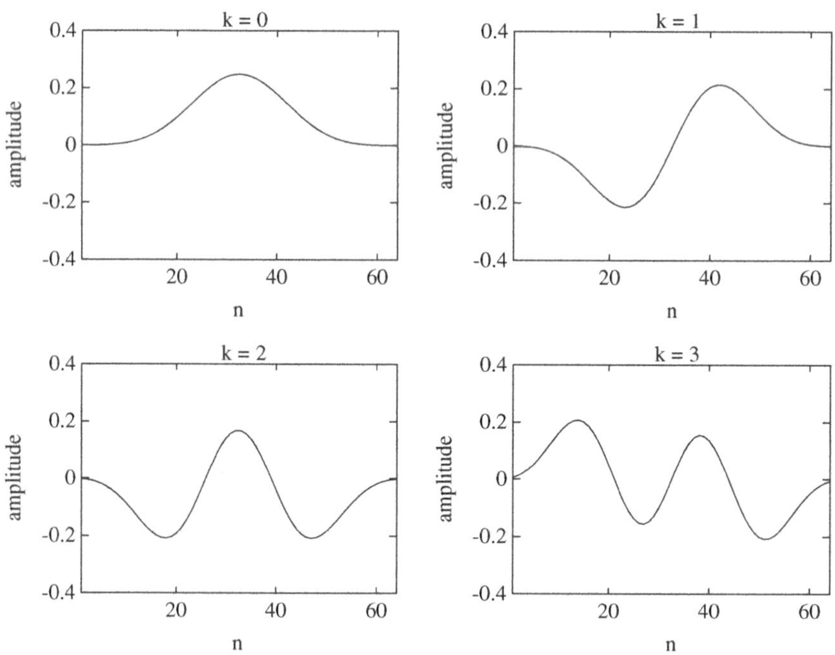

Figure 2.2 The first four data windows for the case $N = 64$ and $NW = 4$. They are simply displays of the first four eigenvectors $\mathbf{v}_{(k)}(N,W)$.

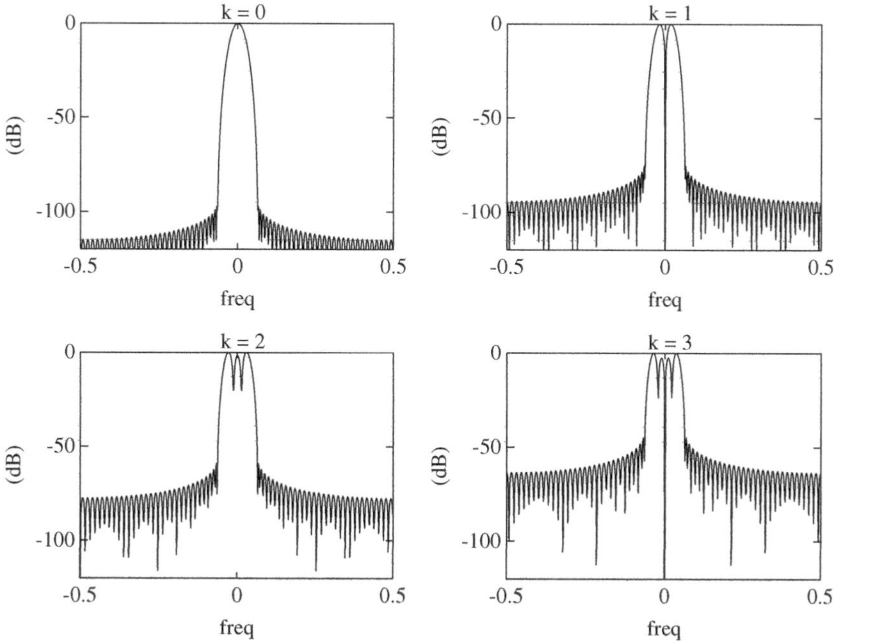

Figure 2.3 The first four spectral windows for the case $N = 64$ and $NW = 4$. These are the complex amplitudes squared (in dB) of the Fourier transforms of the above data windows.

optimum in the sense of energy concentration within the frequency band $(f - W, f + W)$. In essence, by using them, we are maximizing the signal energy within the band $(f - W, f + W)$ and minimizing, at the same time, the energy leakage outside this band. They are therefore the ideal choice to use as a basis of expansion in the frequency domain for band-limited processes. Actually, another way of viewing MTM, is by having the data pass through the baseband (low-pass) filter (Fig. 2.4) as it slides over all frequencies in the interval (–1/2, 1/2). Since spectrum estimation is essentially the estimation of signal power within a certain analysis window and this can ideally be done with a narrow rectangular filter, we see that the baseband filter is the best possible approximation of such a window. The fact that more than one window is used makes for a smaller variance in the estimator. Also, since the signal power concentration within the analysis band is large (eigenvalues close to one), the bias introduced from the multiplicity of windows is kept small.

2.5 TEST DATASET AND A COMPARISON OF SOME POPULAR SPECTRUM ESTIMATION PROCEDURES

In order to check our understanding of the multi-taper method, at each stage of the development, we try to implement it on a known test dataset. This set consists of a complex time series of $N = 64$ points as described in reference 22. It is an extension

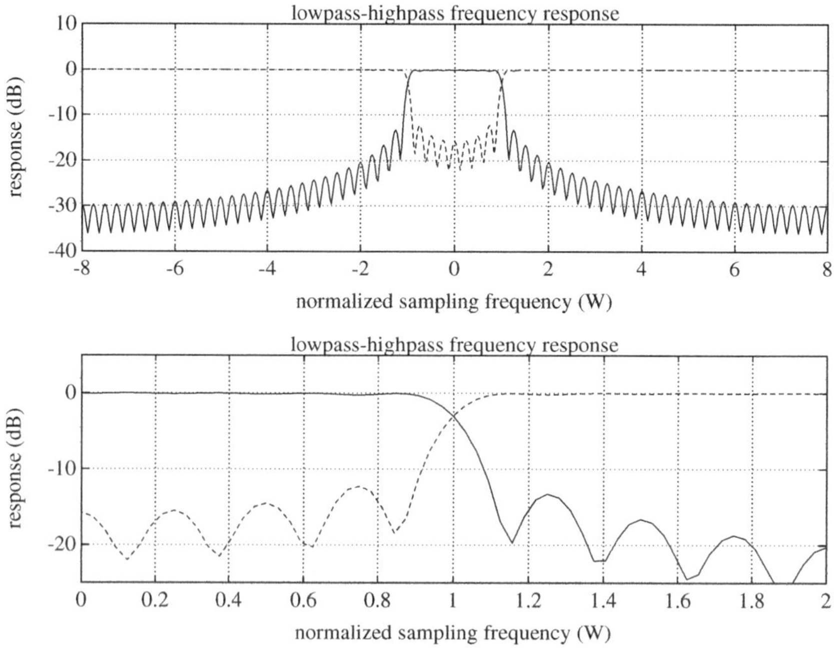

Figure 2.4 Low-pass/high-pass frequency response. The low-pass (baseband) response is the average of the first 8 spectral windows, while the high-pass is the average of the other 56. The low-pass response approximates the ideal rectangular filter for spectrum estimation, while the high-pass one shows the leakage that occurs from outside the analysis window W. If sidelobe reduction is desired (e.g., in filter design), a smaller number of windows should be used. The top part of the figure shows the complete frequency response $-8W \leq f \leq 8W$, scaled in units of the window W, while the lower part expands on the range $0 \leq f \leq 2W$.

to the complex domain of the famous real dataset given in reference 15, where 11 modern methods of spectrum estimation were tested. (As will be seen later on, none of them performed as well as MTM.) The analytic spectrum of this synthetic dataset is composed of the following components:

1. Two complex sinusoids of fractional frequencies 0.2 and 0.21 in order to test the resolution capability of a spectral estimator. Note that the Rayleigh resolution limit is $1/N = 1/64 = 0.015625$, so that the difference of the two fractional frequencies of the doublet is just slightly below it.
2. Two weaker complex sinusoids of 20 dB less power at 0.1 and -0.15. These two singlets were selected to test a spectral estimator's capability to pick out weaker signal components among stronger ones.
3. A colored noise process, generated by passing two independently generated zero-mean real white noise processes through identical moving average filters to separately generate the real and imaginary components of the test data noise process. Each filter has the identical raised cosine response, seen in Fig. 2.6, between fractional frequencies 0.2–0.5 centered at 0.35 or

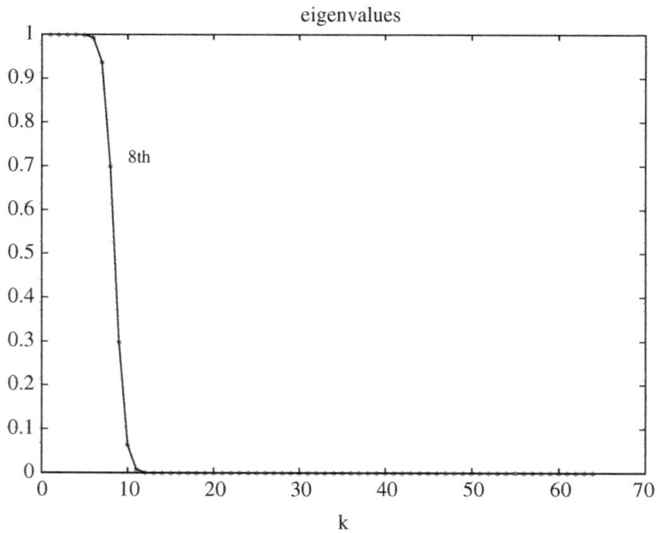

Figure 2.5 The eigenvalue spectrum for $NW = 4$ and $N = 64$. As we can see, the first 8 eigenvalues are very close to 1, corresponding to the first $K = 2NW = 8$ windows that have a negligible effect on the bias of the spectrum estimator.

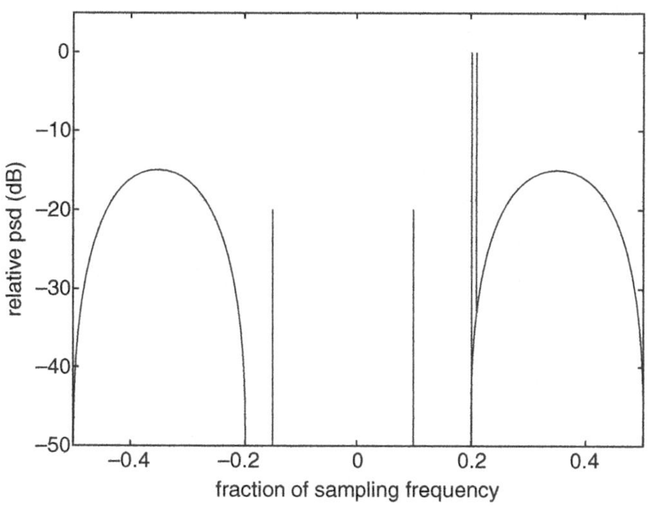

Figure 2.6 The exact known analytic spectrum of Marple's synthetic dataset.

between −0.2 and −0.5 centered at −0.35. The maximum power level of this noise process is 15 dB lower than the doublet and 5 dB higher than the singlets.

Note that even though the shape of the colored noise process is identical in the exact, analytic form of the spectrum for both positive and negative frequencies, this symmetry is not expected to be seen in the estimated spectrum because the real and imaginary components were generated independently.

2.5.1 Classical Spectrum Estimation

Following Marple [22], we first start with the classical spectrum estimation method of simply taking the discrete Fourier transform (implemented with a 4096-point FFT) and a rectangular window. To compare with a case where a window is implemented, the Hamming window (3 segments of 32 samples each, with 16 samples overlap between segments) is then used. Figure 2.7 displays the results, and we can see that some of the spectrum features are picked out already. Of course, since the Fourier transform is essentially a cross-correlation of the data sequence and a complex sinusoid, it tries to fit sinusoids to the continuous part of the spectrum. The Hamming window alleviates the problem to some extent, with a smaller variance on the continuous part of the spectrum, but it results in an increased bias on the line component estimates.

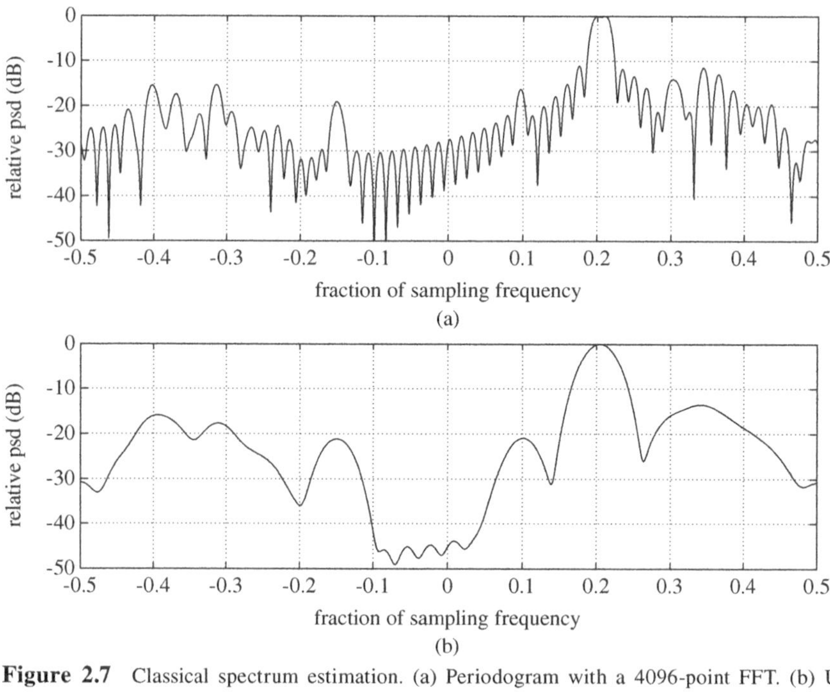

Figure 2.7 Classical spectrum estimation. (a) Periodogram with a 4096-point FFT. (b) Using a Hamming window and a 4096-point FFT.

2.5.2 MUSIC and MFBLP

MUlitple SIgnal Classification (MUSIC) and *Modified Forward Backward Linear Prediction* (MFBLP) are two of the modern algorithms for estimating line components in a data sequence. They both use the concept of a signal and noise subspace. Projection operators are constructed so as to map the data onto one or the other subspace. The signal-space component is thus optimized, affecting a higher SNR that leads to superresolution, provided that the assumptions on which the construction of the projectors is based are valid (i.e., background noise correlation matrix is known, the SNR is above a certain threshold, and the data are properly calibrated).

Choosing the number of signals arbitrarily to be 10, we see (Fig. 2.8) that there is no problem in picking out the line components, but the methods, as expected, try to fit the continuous part of the spectrum with sinusoids as well. Without any *a priori* knowledge, it is easy to mistake a noise peak for a signal peak. We also observe a gradual decay of the eigenvalue spectrum, a clear indication of the existence of colored noise.

To summarize, both the classical and eigendecomposition methods mentioned above, fail in estimating fully and correctly both the line and continuous parts of the given spectrum.

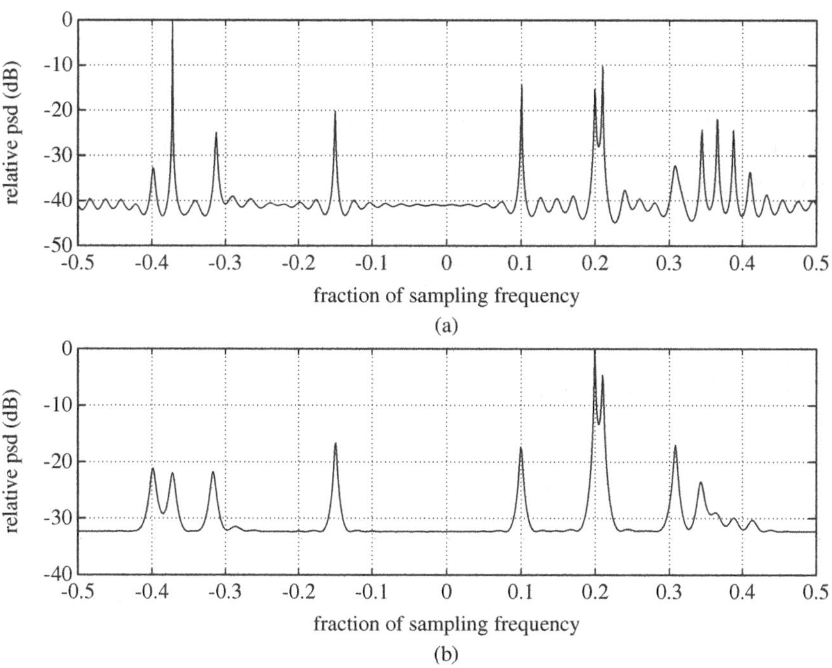

Figure 2.8 MFBLP (a) and MUSIC (b) spectra.

2.6 MULTI-TAPER SPECTRUM ESTIMATION

We now turn our attention to MTM[8] and begin to solve the fundamental equation of spectrum estimation (2.11) by expanding its factors in $(f - W, f + W)$ using the Slepian basis. From *Mercer's theorem*, the kernel expansion is defined by

$$\mathbf{D}_N(f - \nu) = \sum_{k=0}^{\infty} \lambda_k V_k(f) V_k^*(\nu) \tag{2.18}$$

and

$$dZ(f - \nu) = \sum_{k=0}^{\infty} x_k(f) V_k^*(\nu) d\nu \tag{2.19}$$

where the asterisk denotes complex conjugation. Using orthogonality properties carefully (with some help from Yaglom [41]), the coefficients of (2.19) are given by

$$x_k(f) = \sum_{n=0}^{N-1} x(n) \upsilon_n^{(k)}(N, W)^{-j2\pi fn} \tag{2.20}$$

We call the $\{x_k(f)\}$ the *eigencoefficients* of the kth sample. Since they are computed by transforming the data multiplied by the kth data window $\upsilon_n^{(k)}(N, W)$, their absolute squares

$$\hat{S}_k(f) = |x_k(f)|^2 \tag{2.21}$$

are, individually, the direct spectrum estimates, and we therefore call them *eigenspectra*. Using the first $K = 2NW$ terms, the ones with the largest eigenvalues, we obtain a crude multi-taper spectrum estimate as

$$\overline{S}(f) = \frac{1}{K} \sum_{k=0}^{N-1} \frac{1}{\lambda_k(N, W)} |x_k(f)|^2 \tag{2.22}$$

2.6.1 The Adaptive Spectrum

While the lower-order eigenspectra have excellent bias properties, there is some degradation as k increases toward $2NW$. In his 1982 paper [36], Thomson introduces a set of weights $\{d_k(f)\}$ that downweight the higher order eigenspectra. He derives

[8] The name multiple-window or multi-taper is the result of using a multiplicity of windows in the spectrum estimation process instead of just one, as is commonly the practice. This reduces the variance while increasing somewhat the bias in the estimation procedure. However, as long as we are using windows with corresponding eigenvalues $\lambda \approx 1$, the increase in bias is negligible.

them by minimizing the *mean-square error* between $Z_k(f)$, the exact coefficients[9] of the expansion of $dZ(f)$, and $d_k(f)x_k(f)$, which is defined by the expectation

$$\mathbf{E}\left\{|Z_k(f)-d_k(f)x_k(f)|^2\right\} \tag{2.23}$$

Given that

$$Z_k(f)=\frac{1}{\sqrt{\lambda_k}}\int_{-W}^{W}V_k(\nu)\,dZ(f-\nu)$$

and

$$x_k(f)=\int_{-1/2}^{1/2}V_k(\nu)\,dZ(f-\nu)$$

we may write

$$Z_k(f)-d_k(f)x_k(f)=$$
$$\left[\frac{1}{\sqrt{\lambda_k}}-d_k(f)\right]\int_{-W}^{W}V_k(\nu)\,dZ(f-\nu)-d_k(f)⨍V_k(\nu)\,dZ(f-\nu)$$

where the cut integral is defined by

$$⨍=\left(\int_{-1/2}^{1/2}-\int_{-W}^{W}\right)$$

The intervals$(-W, W)$ and $(-1/2, -W) \cup (W, 1/2)$ are named the inner and outer bands, respectively.

It is helpful here to think of the energy in the band$(f - W, f + W)$ as a signal component and the energy outside it as a noise component. These two components are uncorrelated, making the expectation of their cross products zero, and the minimization of (2.23) is then simply a process of finding the *optimum Wiener filter.*

Define the *broadband bias* of the kth eigenspectrum as

$$B_k(f)=\left|⨍V_k(\nu)\,dZ\,(f-\nu)\right|^2 \tag{2.24}$$

The expected value of the broadband bias is

$$\mathbf{E}\{B_k(f)\}=⨍|V_k(\nu)|^2\,S(f-\nu)\,d\nu \tag{2.25}$$

and from the *Cauchy–Schwartz inequality* it can be bounded as

$$\mathbf{E}\{B_k(f)\}\le⨍|V_k(\nu)|^2\,d\nu\,⨍\mathbf{E}\left\{|dZ(f-\nu)|^2\right\} \tag{2.26}$$

The first integral on the right-hand side of (2.26) is just the energy of the Slepian function in the outer band with value$(1 - \lambda_k)$. The second one, by filling the tap in the inner band, has as expected value the average power of the process, σ^2 (the process variance). We therefore write

$$\mathbf{E}\{B_k(f)\}\le(1-\lambda_k)\sigma^2 \tag{2.27}$$

[9] Though unobservable, the exact coefficients of the expansion are important in the sense that they are the expansion coefficients which would be obtained if the entire process were passed through an ideal bandpass filter from $(f - W)$ to $(f + W)$ before truncation to the finite-sample size.

The weights that minimize (2.23) are therefore

$$d_k(f) = \frac{\sqrt{\lambda_k}\, S(f)}{\lambda_k S(f) + \mathbf{E}\{B_k(f)\}} \tag{2.28}$$

The process variance is

$$\sigma^2 = \int_{-1/2}^{1/2} S(f)\, df = r_x(0) = \mathbf{E}\left\{|x(n)^2|\right\} \approx \frac{1}{N}\sum_{n=0}^{N-1} |x(n)|^2 \tag{2.29}$$

A fair initial estimate of the expected value of the broadband bias, $\hat{B}_k(f)$, is given by

$$\hat{B}_k(f) = \mathbf{E}\{B_k(f)\} = (1-\lambda_k)\sigma^2 \tag{2.30}$$

which is actually the upper bound for $\mathbf{E}\{B_k(f)\}$.

Note that in order to compute the adaptive weights $d_k(f)$ in (2.28), we need to know the true spectrum $S(f)$. Of course, if we did, there would be no need to do any spectrum estimation at all. Relation (2.28), however, is useful in setting up an iterative scheme to estimate $S(f)$ as $\hat{S}(f)$, by substituting (2.28) into the estimate

$$\hat{S}(f) = \frac{\sum_{k=0}^{K-1} |d_k(f)|^2\, \hat{S}_k(f)}{\sum_{k=0}^{K-1} |d_k(f)|^2} \tag{2.31}$$

which leads to

$$\sum_{k=0}^{K-1} \frac{\lambda_k\left[\hat{S}(f) - \hat{S}_k(f)\right]}{\left[\lambda_k \hat{S}(f) + \hat{B}_k(f)\right]^2} = 0 \tag{2.32}$$

The solution can be found iteratively from

$$\hat{S}^{(i+1)}(f) = \left[\sum_{k=0}^{K-1} \frac{\lambda_k \hat{S}_k^{(i)}(f)}{\left[\lambda_k \hat{S}^{(i)}(f) + \hat{B}_k(f)\right]^2}\right]\left[\sum_{k=0}^{K-1} \frac{\lambda_k}{\left[\lambda_k \hat{S}^{(i)}(f) + \hat{B}_k(f)\right]^2}\right]^{-1}$$

using as a starting value for $\hat{S}(f)$, the average of the two lowest order eigenspectra. Convergence is usually rapid, with successive spectrum estimates differing by less than 5% in 5–20 iterations.

A simple implementation of the above scheme can be achieved by using (2.30) for $\hat{B}_k(f)$. In his original paper [36], Thomson goes to some length to find a tighter bound than this one. The basic idea is the following: Note that the definition of $\mathbf{E}\{B_k(f)\}$ is a convolution in the frequency domain. Transforming it into the time domain, we only need to multiply two time functions. After doing the multiplication, we can go back into the frequency domain. The whole procedure can be efficiently implemented with the standard fast Fourier transform (FFT) algorithm.

For this purpose, define the outer lag window

$$L_k^{(o)}(\tau) = \oint e^{j2\pi\tau\nu} |V_k(\nu)|^2 \, d\nu = \int_{-1/2}^{1/2} e^{j2\pi\tau\nu} \, |V_k'(\nu)|^2 \, d\nu$$

where

$$V_k'(\nu) = \begin{cases} V_k(\nu) & \text{if } \nu \in (-1/2, W) \cup (W, 1/2) \\ 0 & \text{if } \nu \in (-W, W) \end{cases}$$

The last integral can be approximated with the FFT algorithm.

Next, compute the autocovariance function $R^{(o)}(\tau)$ corresponding to the spectrum estimate of the current iteration:

$$R^{(o)}(\tau) = \int_{-1/2}^{1/2} \hat{S}(\nu) e^{j2\pi\tau\nu} d\nu$$

This integral can also be approximated with the FFT.

Finally, transform back into the frequency domain

$$\hat{B}_k(f) = \sum_{\tau} e^{-j2\pi\tau\nu} L_k^{(o)}(\tau) R^{(o)}(\tau)$$

and this sum can also be computed with another FFT.

In implementing the above idea, a somewhat better resolution is achieved than by using (2.30) at the expense of more computer time. Note, however, that the condition (2.27) is not necessarily satisfied for the first few k. Note also that a good estimate of the autocovariance function can be computed by using the final spectrum estimate above.

A useful byproduct of this adaptive estimation procedure is an estimate of the stability of the estimates, given by

$$\upsilon(f) = 2 \sum_{k=0}^{K-1} |d_k(f)|^2 \tag{2.33}$$

which is the approximate number of degrees of freedom for $\hat{S}(f)$ as a function of frequency. If the average $\bar{\upsilon}$, of $\upsilon(f)$ over frequency, is significantly less than $2K$, then either the window W is too small, or additional prewhitening should be used. This, together with a variance efficiency coefficient that is also developed in reference 36, can provide a useful stopping rule when W and K are varied. In more complicated cases, jackknifed[10] error estimates for the spectrum can be computed as well [40].

[10] The *jackknife*, in the simplest form, refers to the following procedure. Given a set of N observations, each observation is deleted in turn, forming N subsets of $N-1$ observations. These subsets are used to form estimates of a given parameter, which are then combined to give estimates of bias and variance for this parameter, valid under a wide range of parent distributions. Thomson and Chave [40] discuss the extension of this concept to spectra, coherences, and transfer functions.

2.6.2 The Composite Spectrum

The use of adaptive weighting as developed above provides superior protection against leakage and bias. Thomson also offers a further refinement to achieve higher resolution by considering each specific frequency point f_0 as a free parameter in $(f - W \leq f_0 \leq f + W)$. A different choice of weights is the result, leading to the composite spectrum estimate

$$\hat{S}_C(f) = \frac{\int_{f-W}^{f+W} w(f_0)\hat{S}_h(f; f_0)\,df_0}{\int_{f-W}^{f+W} w(f_0)\,df_0} \tag{2.34}$$

where

$$\hat{S}_h(f; f_0) = \frac{2}{\upsilon(f)}\left|\sum_{k=0}^{K-1} V_k(f - f_0)\,d_k(f)\,x_k(f_0)\right|^2 \tag{2.35}$$

$$w(f_0) = \frac{\upsilon(f_0)}{\hat{S}(f_0)^2} \tag{2.36}$$

and $\hat{S}(f_0)$ is the adaptive spectrum developed earlier.

This choice of weights imposes the constraint that $\hat{S}(f_0)$ should have sufficient degrees of freedom for $w(f_0)$ to have a reasonable distribution. In practice, it is inadvisable to do a free parameter expansion over the full window, $|f - f_0| = W$, but rather step near $0.8W$ to $0.9W$ in order to minimize the outside leakage (see Fig. 2.4). Furthermore, in regions where the number of degrees of freedom is small, Thomson suggests rescaling $\hat{S}_h(f;f_0)$ by dividing by a factor proportional to

$$\sum_{k=0}^{K-1} |w_k(f_0)V_k(f - f_0)|^2$$

The implementation of this final version of Thomson's spectrum estimation method was done numerically. Function values of $w(f)$ at any desired frequency point from the already computed data table of $\upsilon(f)$ and $\hat{S}(f)$ were interpolated with splines. If, however, (2.30) is used for $\hat{B}_k(f)$, then it is easy to have an exact expression of $w(f)$ and no interpolation is necessary (the difference in the final composite spectrum estimate between this approach and the one where $w(f)$ is interpolated is almost negligible). The integration was performed numerically with *Romberg's method*,[11] which, for a given numerical accuracy, requires the least number of function evaluations. The integration boundary was chosen to be $0.8W$, because this resulted in line components having a ratio closer to the known one. Finally, it is necessary to explicitly adjust the scaling[12] of $\hat{S}_C(f)$. Note that Thomson's suggestion

[11] Both the spline and Romberg integration subroutines were adapted from reference 30.

[12] That is, multiply by a proper scaling factor in order to get a variance estimate $\hat{\sigma}^2$ (the area under the spectrum computed by trapezoidal integration) close to the known one.

mentioned above on rescaling $\hat{S}_h(f;f_0)$ does not specify the proportionality constant, so there is some justification for this ad hoc rescaling procedure.

A more recent discussion on this *high-resolution* spectrum estimate is given in reference 38. Thomson points out that this is still an area being actively developed; for example, it is shown that although the estimate is unbiased for slowly varying spectra, it underestimates fine spectral structure. In the latter part of this chapter, where Thomson's method is implemented on real data, it is the adaptive spectrum estimator that is used.

2.6.3 Computing the Crude, Adaptive, and Composite Spectra

The results of applying the crude, adaptive, and composite spectra described above are seen in Figs. 2.9 to 2.11. The reason for varying the time–bandwidth product has to do with the bias-variance trade-off that is the result of the ill-posedness of the spectrum estimation problem. If the analysis window W is too small (to better resolve details), we have poor statistical stability (larger variance); but if W is too large, the estimate has poor frequency resolution. We see these effects, both here, in the spectra and in the F-tests considered in the following section. In practice, therefore, we have to try a variety of time–bandwidth product values to pick out the

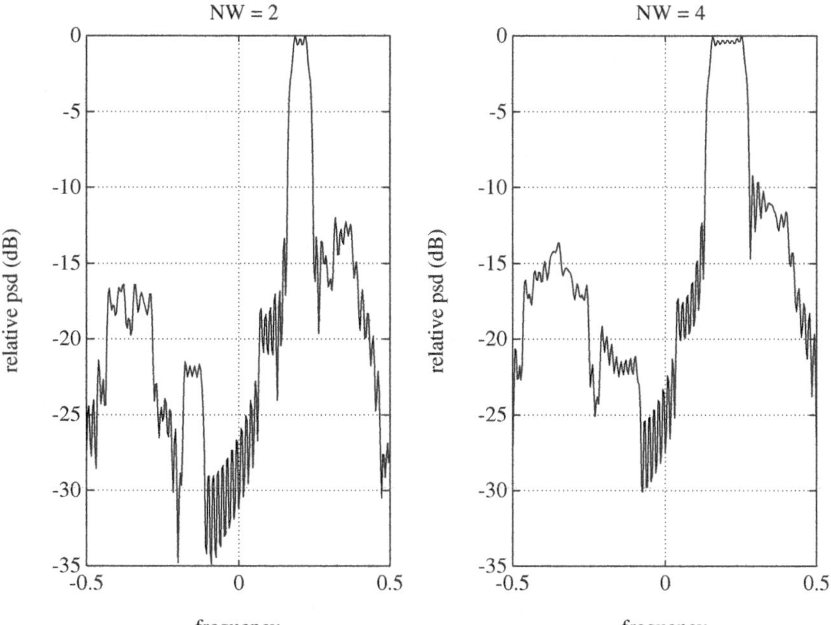

Figure 2.9 The crude spectra $\bar{S}(f)$ for $NW = 2$ and 4.

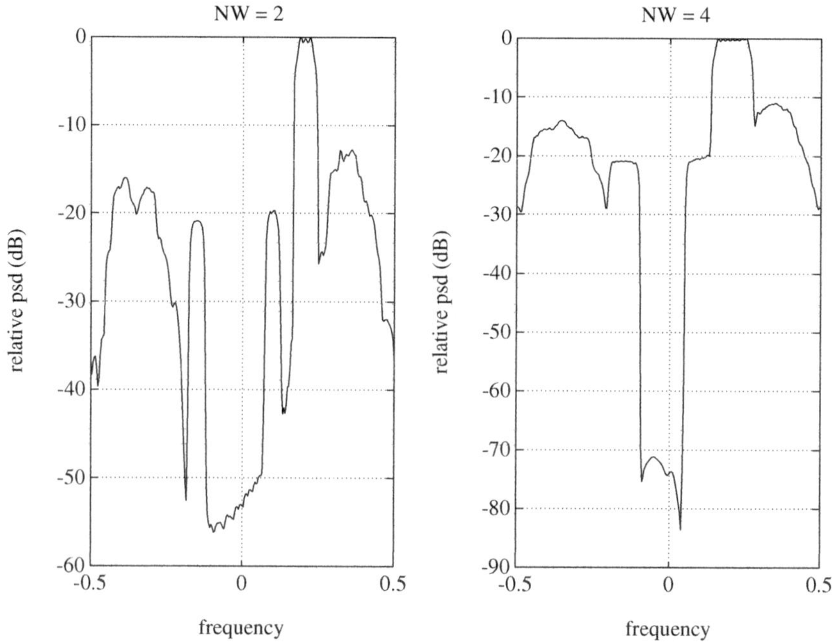

Figure 2.10 The adaptive spectra $\bar{S}(f)$ for $NW = 2$ and 4.

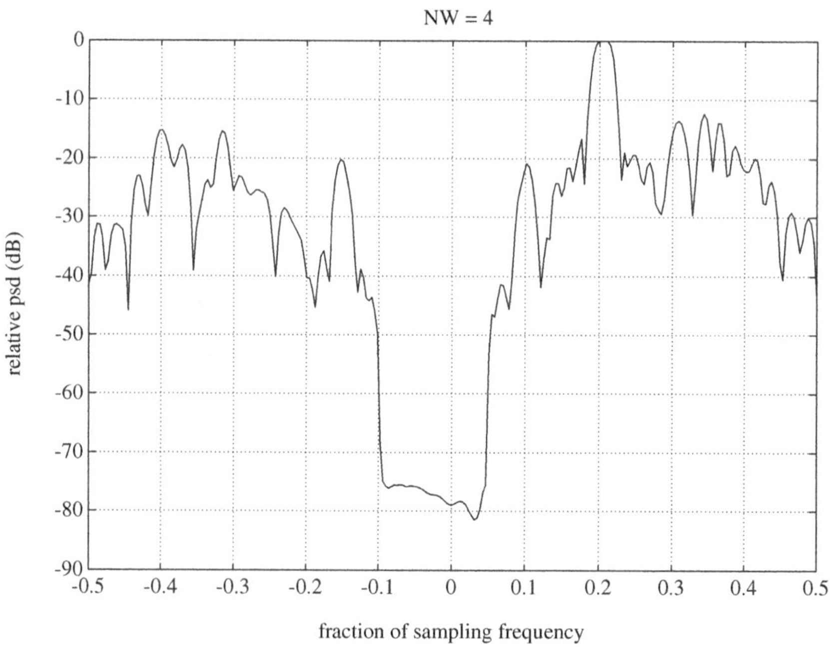

Figure 2.11 The composite spectrum for $NW = 4$, $K = 8$ and an integration boundary of $0.8W$. A larger boundary results in a larger mismatch of the power levels between the known signal components. Note the detail of the composite spectrum despite the larger spectral window.

features that are of interest. Thomson [39] recommends that W be between $1/N$ and $20/N$, with a time–bandwidth product of 4 or 5 being a common starting point.

2.7 *F*-TEST FOR THE LINE COMPONENTS

The spectra computed in Section 2.6.3 are not expected to be good estimates, since we know that line components exist and that the spectrum estimation techniques developed have implicitly assumed none. With MTM, we can apply the *statistical F-test* to check for and estimate existing line components in the spectrum. The *F*-test is a statistical test that assigns a probability value to each of two hypotheses concerning samples taken from a parent population. These samples are assumed to follow a χ^2 distribution, which is the case for an unknown mean and variance sample; it consists of a sum of squares, taken from a Gaussian (normal) population. A brief outline of this test applied to linear regression models, is given in the following subsection; for more details, see Draper and Smith [7].

2.7.1 Brief Outline of the *F*-Test

Let us assume that we have a model described by

$$\mathbf{y} = \mathbf{A}\mathbf{x} + \mathbf{e}$$

that is linear with respect to the $p \times 1$ parameter vector $\mathbf{x}$, where the $n \times p$ coefficient matrix $\mathbf{A}$ and $n \times 1$ vector $\mathbf{y}$ are known or can be estimated from a given dataset. We assume that the error vector $\mathbf{e}$ has independent components that come from $N(0,\sigma^2)$. Therefore, another way to express our assumed model is to write

$$\mathbf{E}\{\mathbf{y}\} = \mathbf{A}\mathbf{x}$$

In order to get the best possible estimate of our parameter vector $\mathbf{x}$ in the least-squares sense, we have to find

$$\min_{\mathbf{x}} \|\mathbf{y} - \mathbf{A}\mathbf{x}\|^2$$

Using the superscript H to denote the Hermitian transposition of a matrix, we may express the squared error as

$$\mathbf{e}^2(\mathbf{x}) = \mathbf{e}^H\mathbf{e} = \|\mathbf{y} - \mathbf{A}\mathbf{x}\|^2 = \mathbf{y}^H\mathbf{y} - \mathbf{y}^H\mathbf{A}\mathbf{x} - \mathbf{x}^H\mathbf{A}^H\mathbf{y} + \mathbf{x}^H\mathbf{A}^H\mathbf{A}\mathbf{x}$$

which assumes its minimum value at the well-known *linear least-squares solution*

$$\hat{\mathbf{x}} = \left(\mathbf{A}^H\mathbf{A}\right)^{-1}\mathbf{A}^H\mathbf{y} = \mathbf{A}^+\mathbf{y}$$

where $\mathbf{A}^+ = (\mathbf{A}^H\mathbf{A})^{-1}\mathbf{A}^H$ is the pseudo-inverse of $\mathbf{A}$.

The *F*-test comes about from observing that we can break the observed total variance $\mathbf{y}^H\mathbf{y}$ of our model into two components, one due to the regression itself, $\|\mathbf{A}\hat{\mathbf{x}}\|^2$ [7, p. 80], and the other, $\|y - \mathbf{A}\hat{\mathbf{x}}\|^2$, due to residual errors. Each of these components has associated with it a number of degrees of freedom, ν_1 and ν_2, respectively. For K complex data points, the total number of degrees of freedom is

$$\nu_1 + \nu_2 = 2K$$

It turns out now that, provided the errors are independent and zero-mean Gaussian random variables, each of the two variance components follows a χ^2-distribution with ν_1 or ν_2 degrees of freedom, respectively. Their ratio follows the $F(\nu_1,\nu_2)$ distribution, and we can make hypothesis testing at a desired level of significance. Table 2.1 shows this simple *analysis of variance (ANOVA) breakdown.*

We test the hypothesis H_0: $\mathbf{x} = 0$ against H_1: $\mathbf{x} \neq 0$ at a significance level α as follows: If H_0 is true, then the ratio (see Table 2.1)

$$F = \frac{MS_{reg}}{s^2} = \frac{\nu_2 SS_1}{\nu_1 SS_2}$$

should follow the $F(\nu_1,\nu_2)$ distribution whose value for a significance level of α is found from statistical tables. If the computed ratio is larger than the table value,[13] then our hypothesis is rejected with $100(1 - \alpha)\%$ confidence. This means that at least one of the components of $\mathbf{x}$ is different from zero.

It is also possible to test various linear hypotheses about the model under consideration. Focusing on the possible addition of extra parameters, consider the two models:

1. $\mathbf{E}\{Y\} = A_1x_1 + A_2x_2 + \ldots + A_px_p$
2. $\mathbf{E}\{Y\} = A_1x_1 + A_2x_2 + \ldots + A_qx_q$

where $q < p$. The A's in both models are the same when the subscripts are the same. We simply have fewer parameters to fit in the second one compared to the first. From the first model, we estimate $\mathbf{x}$ as

$$\hat{\mathbf{x}}_p = \left(\mathbf{A}_p^H \mathbf{A}_p\right)^{-1} \mathbf{A}_p^H \mathbf{y}$$

The corresponding residual sum of squares is

$$S_1 = \|\mathbf{y} - \mathbf{A}_p \hat{\mathbf{x}}_p\|^2$$

This has $2(n - p)$ degrees of freedom (n is the total number of complex data points we have available for the regression analysis), and

Table 2.1 The Basic ANOVA Table

Variation Source	Degrees of Freedom	Sum of Squares (SS)	Mean SS
Regression	ν_1	$SS_1 = \|\mathbf{A}\hat{\mathbf{x}}\|^2$	$MS_{\text{reg}} = SS_1/\nu_1$
Residuals	ν_2	$SS_2 = \|y - \mathbf{A}\hat{\mathbf{x}}\|^2$	$s^2 = SS_2/\nu_2$

[13] In practice, of course, we require a computed F ratio that is much larger than the tabulated one.

$$s^2 = \frac{S_1}{2(n-p)}$$

is also an estimate for the squared variance for model 1. For the second model, we have respectively

$$\hat{\mathbf{x}}_q = (\mathbf{A}_q^H \mathbf{A}_q)^{-1} \mathbf{A}_q^H \mathbf{y}$$

and

$$S_2 = \|\mathbf{y} - \mathbf{A}_q \hat{\mathbf{x}}_q\|^2$$

where S_2 has $2(n-q)$ degrees of freedom. To test the hypothesis that the extra terms are unnecessary, we consider the ratio

$$\frac{\frac{1}{2(p-q)}(S_2 - S_1)}{\frac{1}{2(n-p)} S_1}$$

and refer it to the $F[2(p-q), 2(n-p)]$ distribution in the usual manner.

This provides us with enough information to set up partial F-tests in the following subsections, for computing multiple spectral lines.

2.7.2 The Point Regression Single-Line *F*-Test

Given a line component at frequency f_0, the expected value of the kth eigencoefficient is given by

$$\mathbf{E}\{x_k(f)\} = \mu V_k(f - f_0) \tag{2.37}$$

At a given frequency f, the $x_k(f)$'s correspond to $\mathbf{y}$ in subsection 2.7.1, the parameter μ corresponds to $\mathbf{x}$, and the $V_k(f - f_0)$'s correspond to matrix $\mathbf{A}$. The number of parameters p is equal to 1 here, and by setting it up as a least-squares problem, we have to minimize the residual error $\|\mathbf{x} - \mathbf{V}\mu\|^2$ with respect to the 1×1 complex scalar parameter μ, where

$$\mathbf{x} = \begin{bmatrix} x_0(f) \\ x_1(f) \\ \vdots \\ x_{K-1}(f) \end{bmatrix}, \qquad \mu = \mu, \qquad \mathbf{V} = \begin{bmatrix} V_0(f-f_0) \\ V_1(f-f_0) \\ \vdots \\ V_{K-1}(f-f_0) \end{bmatrix}$$

The term *point regression* comes about because we consider each frequency point f as a possible candidate for f_0. We set $f_0 = f$ and test the model (2.37) for statistical significance.

The residual error at f can also be written as

$$e^2(\mu, f) = \sum_{k=0}^{K-1} |x_k(f) - \mu(f) V_k(0)|^2 \tag{2.38}$$

and the least-squares solution of μ is

$$\hat{\mu} = \left(\mathbf{V}^H\mathbf{V}\right)^{-1}\mathbf{V}^H\mathbf{x} = \hat{\mu}(f) = \frac{\sum_{k=0}^{K-1} V_k^*(0)\,x_k(f)}{\sum_{k=0}^{K-1} |V_k(0)|^2}$$

which may also be written as

$$\hat{\mu}(f) = \sum_{n=0}^{N-1} h_n(N,W)\,x(n)\,e^{-j2\pi fn} \tag{2.39}$$

where the nth it harmonic data window is defined by

$$h_n(N,W) = \frac{\sum_{k=0}^{K-1} V_k^*(0)\,\upsilon_n^{(k)}(N,W)}{\sum_{k=0}^{K-1} |V_k(0)|^2} \tag{2.40}$$

We can now test the hypothesis H_0 that when $f_0 = f$, the model (2.37) is false; that is, the parameter μ is equal to zero, versus the hypothesis H_1, where it is not. In other words, the ratio of the energy explained by the assumption of a line component at the given frequency, to the residual energy, gives us the F-ratio as explained in the previous subsection, namely,

$$F(f) = \frac{\frac{1}{\nu_1}\|\mathbf{V}\mu\|^2}{\frac{1}{\nu_2}\|\mathbf{x} - \mathbf{V}\mu\|^2}$$

or

$$F(f) = \frac{\frac{1}{\nu}|\hat{\mu}(f)|^2 \sum_{k=0}^{K-1} |V_k(0)|^2}{\left(\frac{1}{2K-\nu}\right) e^2(\hat{\mu} - f)} \tag{2.41}$$

where ν is equal to two degrees of freedom (real and imaginary parts of the complex line amplitude). The total number $2K$ of degrees of freedom come about from the K complex data points we have available to draw information from. If F is large at a certain frequency, then the hypothesis is rejected; that is, a line component does exist there. The location of the maximum of F provides an estimate of the line frequency that has a resolution within 5–10% of the Cramér–Rao bound.

The test works well if the lines considered are isolated—that is, if there is only a single line in the interval $(f - W, f + W)$. The total number of lines is not important as long as they occur singly. For lines spaced closer than W, we may use a multiple-line test, which is similar but algebraically a more complicated regression of the eigencoefficients on a matrix of functions $V_k(f - f_i)$ with the simple F-test replaced by partial F-tests. Thomson notes that multiple-line tests should be used with caution, because the Cramér–Rao bounds for line parameter estimation degrade rapidly when the line spacing becomes less than $2/N$.

Note also that the F-test is a statistical test. This means that given a large number of different realizations of a data sequence, highly significant values can sometimes be seen, which, in reality, are only sampling fluctuations. A good rule-of-thumb, as Thomson [36] points out, is not to get excited by significance levels below $1 - 1/N$. Experience also suggests to try the test for a variety of NW values. Line components that disappear from one case to the other are almost certainly sampling fluctuations.

In Fig. 2.12, we see the results of applying this F-test to the Marple dataset. Note the effect of the different values of NW and K. The 99% and 95% confidence levels are drawn as well.

2.7.3 The Integral Regression Single-Line *F*-Test

Thomson [36] also suggests an integral regression test instead of the point regression at f_0 that was developed above. The criterion here is the minimization of the sum of integrals

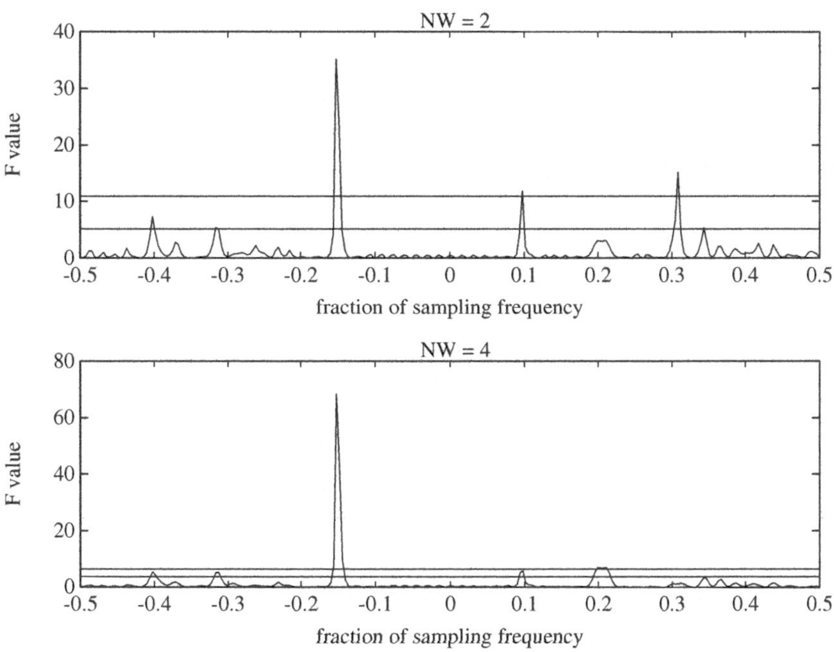

Figure 2.12 The single-line point regression F-test for $NW = 2$ and 4, applied to the Marple dataset. The corresponding F values for 99% and 95% confidence levels are drawn as well.

$$e^2 = \sum_{k=0}^{K-1} \int_{f_0-W}^{f_0+W} |x_k(f) - \mu V_k(f - f_0)|^2 \, df \tag{2.42}$$

with respect to μ. The development is more complex now, but the underlying logic is the same as before. Again, we have an equivalent harmonic window, but here it is formed of convolutions of prolate functions that have exceedingly low sidelobes. The drawback is that as it uses information from a wider bandwidth, it is therefore subject to noise from the same bandwidth. No further details are given for this approach or the multiple line *F*-test in reference 36, but the matrix formalism given earlier is a general enough framework that should be able to handle this situation.[14]

Differentiating the squared error e^2 with respect to μ and setting the result equal to zero yields

$$\int_{f_0-W}^{f_0+W} \left(\mathbf{V}^H \mathbf{V} \mu - \mathbf{V}^H \mathbf{x}\right) df = 0 \tag{2.43}$$

and the 1 × 1 complex parameter $\hat{\mu}$(scalar for this single-line case) is given by

$$\hat{\mu}(f_0) = \frac{\displaystyle\sum_{k=0}^{K-1} \int_{f_0-W}^{f_0+W} V_k^*(f - f_0)\, x_k(f)\, df}{\displaystyle\sum_{k=0}^{K-1} \int_{f_0-W}^{f_0+W} |V_k(f - f_0)|^2 \, df} \tag{2.44}$$

From known formulas, repeated here for convenience, we have [see (2.20) and (2.17), respectively]

$$x_k(f) = \sum_{n=0}^{N-1} x(n)\, \upsilon_n^{(k)}(N, W)\, e^{-j2\pi fn} \quad \text{and} \quad V_k(f) = \sum_{n=0}^{N-1} \upsilon_n^{(k)}(N, W)\, e^{-j2\pi fn}$$

and the Toeplitz matrix eigenvalue equation,

$$\sum_{m=0}^{N-1} \left(\frac{\sin 2\pi W(n-m)}{\pi(n-m)} \right) \upsilon_m^{(k)} = \lambda_k \upsilon_n^{(k)}$$

We do the algebra required for each term in (2.44) and finally end up with the denominator of that equation having the form

$$\sum_{k=0}^{K-1} \int_{f_0-W}^{f_0+W} |V_k(f - f_0)|^2 \, df = \sum_{k=0}^{K-1} \lambda_k \tag{2.45}$$

while the numerator of the equation is

$$\sum_{k=0}^{K-1} \int_{f_0-W}^{f_0+W} V_k^*(f - f_0)\, x_k(f)\, df = \sum_{n=0}^{N-1} \sum_{k=0}^{K-1} \lambda_k \left[\upsilon_n^{(k)}\right]^2 x(n)\, e^{-j2\pi f_0 n} \tag{2.46}$$

[14] In his more recent work [38], Thomson discusses the *F*-tests, both single and multiple, in some detail.

These two relations simplify (2.44) into the form

$$\hat{\mu}(f_0) = \sum_{n=0}^{N-1} h_n x(n) e^{-j2\pi f_0 n}$$

where the harmonic windows now are defined by

$$h_n = \frac{\sum_{k=0}^{K-1} \lambda_k \left[\upsilon_n^{(k)}\right]^2}{\sum_{k=0}^{K-1} \lambda_k}$$

Substituting $\hat{\mu}(f_0)$ for μ into (2.42) for the squared error e^2, and intercharging the order of summation and integration, we obtain

$$e^2(\hat{\mu}, f_0) = \int_{f_0-W}^{f_0+W} \left[\sum_{k=0}^{K-1} |x_k(f)|^2 - |\hat{\mu}|^2 \sum_{k=0}^{K-1} |V_k(f-f_0)|^2 \right] df$$

The first term on the right-hand side of this equation gives us

$$\begin{aligned} Q &= \sum_{k=0}^{K-1} \int_{f_0-W}^{f_0+W} |x_k(f)|^2 \, df \\ &= \sum_{k=0}^{K-1} \sum_{n=0}^{N-1} \sum_{m=0}^{N-1} x(n) x^*(m) \upsilon_n^{(k)} \upsilon_m^{(k)} \left(\frac{\sin[2\pi(n-m)W]}{\pi(n-m)} \right) e^{-j2\pi(n-m)f_0} \end{aligned}$$

and a brute-force computation of this triple sum for each frequency point may take a few hours of CPU time, depending, of course, on the number of frequency points in question. Looking into the symmetries involved and after some algebra, we finally end up with the expression

$$\begin{aligned} Q = {} & 2W \sum_{k=0}^{K-1} \sum_{n=0}^{N-1} \left| x(n) \upsilon_n^{(k)} \right|^2 + \\ & 2\,\mathbf{Re} \left\{ \sum_{k=0}^{K-1} \sum_{n=0}^{N-2} x(n) \upsilon_n^{(k)} \sum_{l=1}^{N-n-1} x^*(l+n) \upsilon_{l+n}^{(k)} \left(\frac{\sin 2\pi l W}{\pi l} \right) e^{j2\pi l f_0} \right\} \end{aligned} \tag{2.47}$$

where **Re** {•} defines the real part of the complex quantity contained inside the braces. Some judicious precomputing, particularly of the exponentials, can drop the CPU time for this sum to only a few minutes and our estimate of the squared error becomes

$$e^2(\hat{\mu}, f_0) = Q - |\hat{\mu}|^2 \sum_{k=0}^{K-1} \lambda_k \tag{2.48}$$

Another, perhaps more straightforward way to compute Q faster is to perform the summation with respect to k first, thus creating a data window matrix and reducing the triple sum to a double one. In complete analogy with the point regression test, the F ratio is now defined by

$$F(f_0) = \frac{\frac{1}{\nu}|\hat{\mu}(f_0)|^2 \sum_{k=0}^{K-1} \lambda_k}{\frac{1}{2K-\nu} e^2(\hat{\mu}, f_0)} \tag{2.49}$$

The method should give similar results as compared to the point regression single-line F-test, and Fig. 2.17 (presented at the end of the next subsection) shows that this is indeed the case.

2.7.4 The Point Regression Double-Line *F*-Test

The model we now have is described by

$$\mathbf{E}\{x_k(f)\} = \mu_1 V_k(f - f_1) + \mu_2 V_k(f - f_2) \tag{2.50}$$

We want to minimize the least-squares residual error

$$e^2 = \|\mathbf{x} - \mathbf{V}\boldsymbol{\mu}\|^2$$

where now

$$\mathbf{x} = \begin{bmatrix} x_0(f) \\ x_1(f) \\ \vdots \\ x_{K-1}(f) \end{bmatrix}, \quad \boldsymbol{\mu} = \begin{bmatrix} \mu_1 \\ \mu_2 \end{bmatrix}, \quad \mathbf{V} = \begin{bmatrix} V_0(f - f_1) & V_0(f - f_2) \\ V_1(f - f_1) & V_1(f - f_2) \\ \vdots & \vdots \\ V_{K-1}(f - f_1) & V_{K-1}(f - f_2) \end{bmatrix} \tag{2.51}$$

If we set

$$\mathbf{V}^H \mathbf{V} = \begin{bmatrix} d_1 & c \\ c^* & d_2 \end{bmatrix} \tag{2.52}$$

then the least-squares solution is

$$\hat{\boldsymbol{\mu}} = \left(\mathbf{V}^H \mathbf{V}\right)^{-1} \mathbf{V}^H \mathbf{x} = \frac{1}{d_1 d_2 - |c|^2} \begin{bmatrix} d_2 & -c \\ -c^* & d_1 \end{bmatrix} \begin{bmatrix} c_1 \\ c_2 \end{bmatrix} \tag{2.53}$$

where

$$d_1 = \sum_{k=0}^{K-1} |V_k(f - f_1)|^2, \qquad d_2 = \sum_{k=0}^{K-1} |V_k(f - f_2)|^2$$

$$c = \sum_{k=0}^{K-1} V_k^*(f - f_1) V_k(f - f_2)$$

$$c_1 = \sum_{k=0}^{K-1} V_k^*(f - f_1), \qquad c_2 = \sum_{k=0}^{K-1} V_k^*(f - f_2) x_k(f)$$

and the *F*-test, for testing the hypothesis that at a given frequency we need at most two lines to explain the data, becomes

$$F(f; f_1, f_2) = \frac{1/\nu}{1/(2K-\nu)} \left[\frac{\|\mathbf{V}\boldsymbol{\mu}\|^2_{\text{double}}}{\|\mathbf{x} - \mathbf{V}\boldsymbol{\mu}\|^2_{\text{double}}} \right] \tag{2.54}$$

Here we have added two degrees of freedom for the extra μ parameter, making $\nu = 4$. This tests the hypothesis that at a certain frequency pair we have one- or two-line components. The modification to the hypothesis that we have two-line components only is expressed as

$$F(f; f_1, f_2) = \frac{1/(4-2)}{1(2K-4)} \left[\frac{\|\mathbf{V}\boldsymbol{\mu}\|^2_{\text{double}} - \|\mathbf{V}\boldsymbol{\mu}\|^2_{\text{single}}}{\|\mathbf{x} - \mathbf{V}\boldsymbol{\mu}\|^2_{\text{double}}} \right] \tag{2.55}$$

For the single-line model, parameter μ is given by

$$\mu_{\text{single}} = c_1 / d_1$$

and

$$\|\mathbf{V}\boldsymbol{\mu}\|^2_{\text{single}} = |c_1|^2 / d_1$$

The introduction of the above single-line model term changes only the scale of the function *F*, if anything for the worst, without affecting the signal peak locations or μ estimates of the double-line model. Note also that the double-line model assumes two lines—that is, $f_1 \neq f_2$. If this condition does not hold, then the denominator in the μ estimates could be zero. In implementing the computer program, care must be taken that this case is excluded.

If we now let $f_1 = f$ and $\Delta f = f_1 - f_2$ vary within $[-W, W]$, we have a two-dimensional surface for *F* that should be capable of resolving two lines that are spaced close together within the window bandwidth *W*. Figures 2.13 to 2.16, computed for a time–bandwidth product $NW = 2$ and 4, illustrate that this is indeed the case.

Zooming around the area of interest, we see that the doublet peaks line up at 0.2 and 0.21 (where we know they should). The fact that the surface plot displays two peaks is due to our mode of representation of using f and Δf as the independent variables. The two peaks are at $(f, f - \Delta f)$, and for $\Delta f = -0.01$ this corresponds to (f_1, f_2), while for $\Delta f = 0.01$ it corresponds to (f_2, f_1) where $f_1 = 0.2$ and $f_2 = 0.21$. The first pair is consistent with our model and has a larger *F* value, so it is the one we pick. The advantage with this particular representation is that we can project the maximum of $F(f, \Delta f)$ onto the f-axis and resolve the doublet from a simpler, one-dimensional function.

Note, however, the appearance of spurious peaks, particularly near the edges of the window boundary. These are explained by the fact that the sliding window (Fig. 2.4) is not an ideal bandpass filter; hence, energy from outside the window, particularly near the window boundaries, can affect the estimation inside the window.

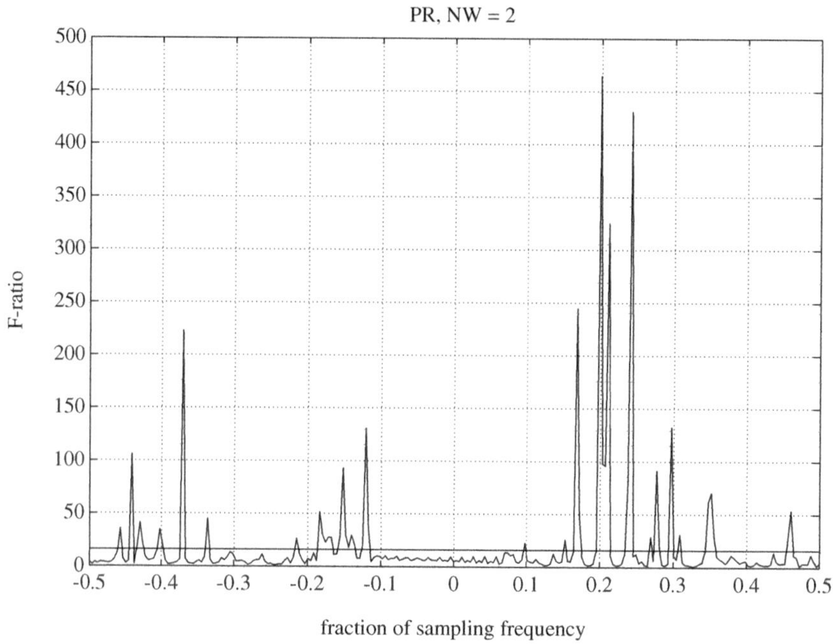

Figure 2.13 The projection of max $F(f,\Delta f)$, onto the f axis. The 99% confidence level is also drawn and we see some spurious peaks above it. As the surface plot in Fig. 2.14 shows, the largest peaks are mostly due to leakage outside the window.

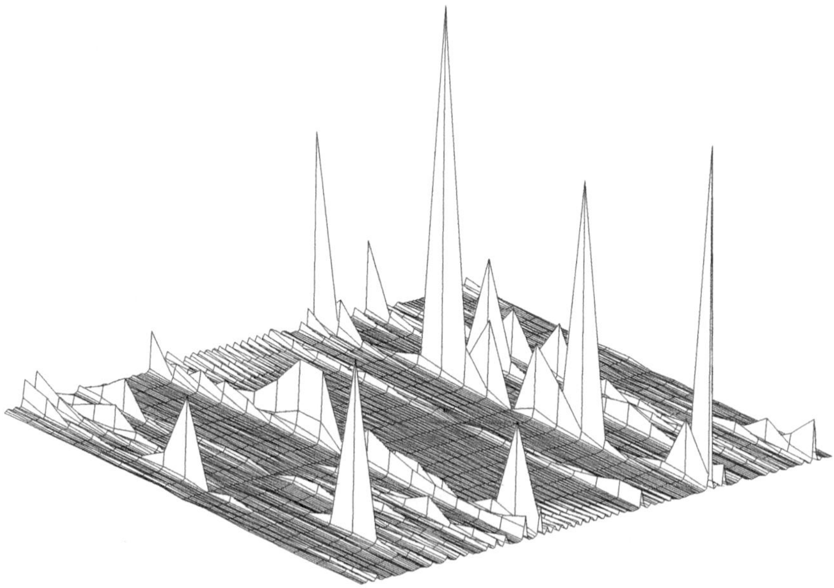

Figure 2.14 The double-line point regression F-test for $NW = 2$. The surface $F(f,\Delta f)$ is shown with a grid size $\frac{1}{256}$ due to the 256-FFTs used. Note the large spurious peaks at the window boundary.

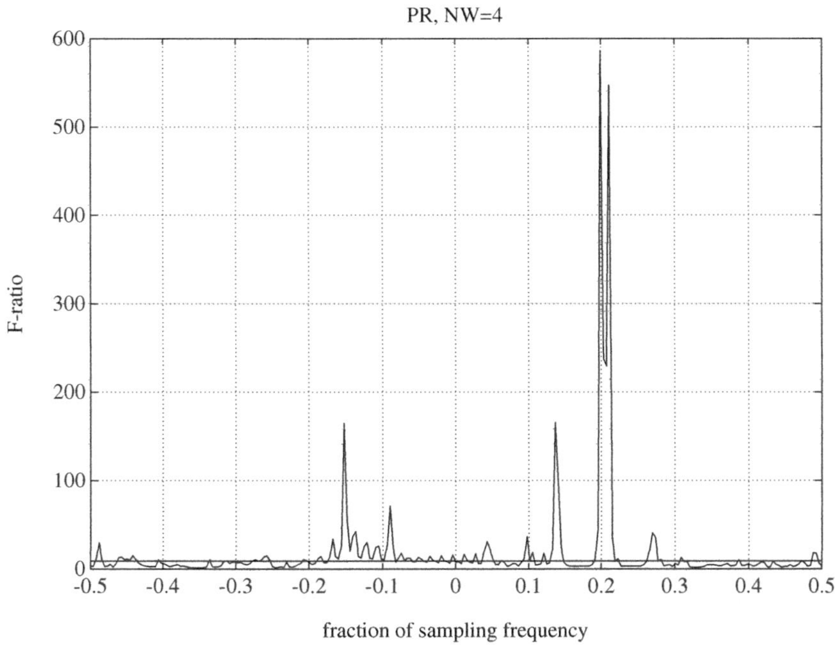

Figure 2.15 The projection of max $F(f,\Delta f)$, onto the f axis.

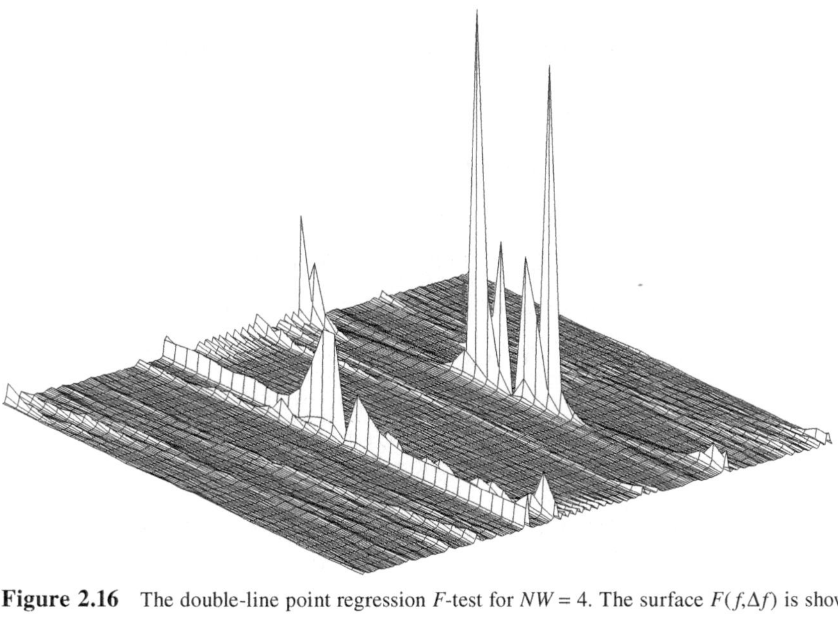

Figure 2.16 The double-line point regression F-test for $NW = 4$. The surface $F(f,\Delta f)$ is shown.

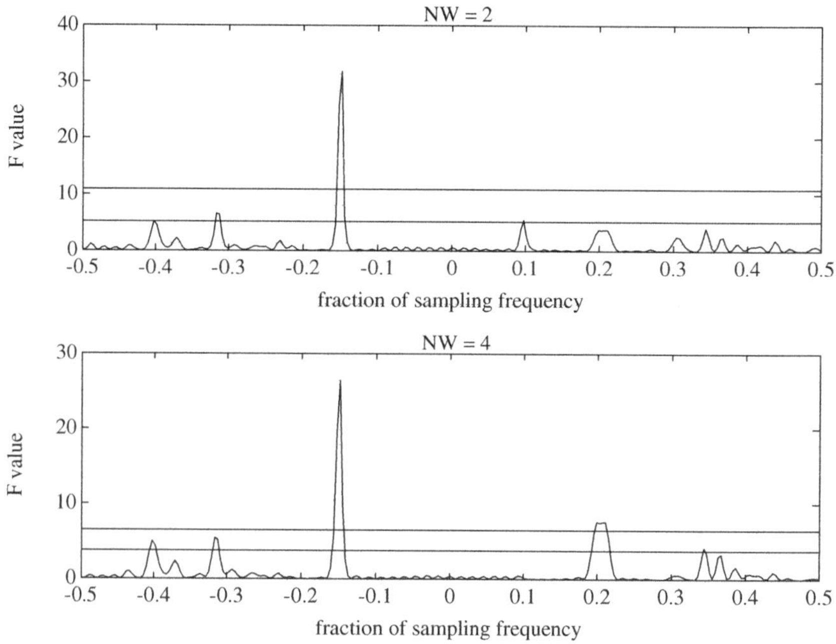

Figure 2.17 The single-line integral regression F-test for $NW = 2$ and 4.

2.7.5 The Integral Regression Double-Line *F*-Test

The model we consider is again

$$\mathbf{E}\{x_k(f)\} = \mu_1 V_k(f - f_1) + \mu_2 V_k(f - f_2)$$

but the minimization criterion now becomes

$$\min_{\mu} e^2(f_1, f_2) = \min_{\mu} \int_{f_1 - W}^{f_1 + W} \|\mathbf{x} - \mathbf{V}\boldsymbol{\mu}\|^2 \, df \tag{2.56}$$

Differentiating e^2 with respect to $\boldsymbol{\mu}$ and setting the result equal to zero yields

$$\int_{f_1 - W}^{f_1 + W} (\mathbf{V}^H \mathbf{V}\boldsymbol{\mu} - \mathbf{V}^H \mathbf{x}) \, df = 0$$

where the integration is performed directly on each term of the matrices or vectors involved. It is assumed that f_2 is within $(f_1 - W, f_1 + W)$. A detailed expression for each component is given in the previous section. The least-squares solution for $\boldsymbol{\mu}$ is

$$\hat{\boldsymbol{\mu}}(f_1, f_2) = \int_{f_1 - W}^{f_1 + W} (\mathbf{V}^H \mathbf{V})^{-1} \mathbf{V}^H \mathbf{x} \, df = \frac{1}{\delta_1 \delta_2 - |\gamma|^2} \begin{bmatrix} \delta_2 & -\gamma \\ -\gamma^* & \delta_1 \end{bmatrix} \begin{bmatrix} \gamma_1 \\ \gamma_2 \end{bmatrix} \tag{2.57}$$

where

$$\delta_1 = \int_{f_1-W}^{f_1+W} d_1 df = \ldots = \sum_{k=0}^{K-1} \lambda_k$$

$$\delta_2 = \int_{f_1-W}^{f_1+W} d_2 df = \ldots = \sum_{k=0}^{K-1} \sum_{n=0}^{N-1} \sum_{m=0}^{N-1} \upsilon_n^{(k)} \upsilon_m^{(k)} e^{j2\pi(n-m)(f_2-f_1)} \frac{\sin 2\pi(n-m)W}{\pi(n-m)}$$

$$\gamma = \int_{f_1-W}^{f_1+W} c df = \ldots = \sum_{n=0}^{N-1} \sum_{k=0}^{K-1} \lambda_k \left[\upsilon_n^{(k)}\right]^2 e^{j2\pi n(f_2-f_1)}$$

$$\gamma_1 = \int_{f_1-W}^{f_1+W} c_1 df = \ldots = \sum_{n=0}^{N-1} \sum_{k=0}^{K-1} \lambda_k \left[\upsilon_n^{(k)}\right]^2 x(n) e^{-j2\pi n f_1}$$

$$\gamma_2 = \int_{f_1-W}^{f_1+W} c_2 df = \ldots = \sum_{k=0}^{K-1} \sum_{n=0}^{N-1} \sum_{m=0}^{N-1} \upsilon_n^{(k)} \upsilon_m^{(k)} x(m) e^{j2[\pi(n-m)f_1 - nf_2]} \frac{\sin 2\pi(n-m)W}{\pi(n-m)}$$

and (2.57) becomes

$$\hat{\mu}(f_1, f_2) = \frac{1}{\delta_1\delta_2 - |\gamma|^2} \begin{bmatrix} \delta_2\gamma_1 - \gamma\gamma_2 \\ -\gamma^* \gamma_1 + \delta_1\gamma_2 \end{bmatrix} \tag{2.58}$$

We also have Q from (2.47). We just have to set $f_0 = f_1$. The regression error term for the double-line model is therefore

$$\zeta = \int_{f_1-W}^{f_1+W} \|\mathbf{V}\boldsymbol{\mu}\|^2_{\text{double}} df = \delta_1 |\mu_1|^2 + \delta_2 |\mu_2|^2 + 2\,\mathbf{Re}\left\{\gamma\, \mu_1^* \mu_2\right\}$$

On the other hand, for the single-line model, we have

$$\int_{f_1-W}^{f_1+W} \|\mathbf{V}\mu\|^2_{\text{single}} df = |\gamma_1|^2 / \delta_1$$

We finally obtain for our F-test the expression

$$F(f_1, f_2) = \frac{1/4}{1/(2K-4)} \left[\frac{\zeta}{Q-\zeta}\right] \tag{2.59}$$

or

$$F(f_1, f_2) = \frac{1/(4-2)}{1/(2K-4)} \left[\frac{\zeta - |\gamma_1|^2/\delta_1}{Q-\zeta}\right] \tag{2.60}$$

Equation (2.59) tests the existence of one or two lines, whereas (2.60) tests the existence of two lines only.

If again, we let f_1 vary over the entire frequency range, f_2 will vary in $(f_1 - W, f_1 + W)$. As we can see from Figs. 2.18 and 2.19 for $NW = 2$, the doublet is resolved and the spurious peak problem disappears. Figure 2.20 shows the corresponding result for $NW = 4$.

2.7.6 Line Component Extraction

The next step in the spectrum estimation procedure is to extract the line components in order to be left with only the continuous part of the spectrum under investigation.

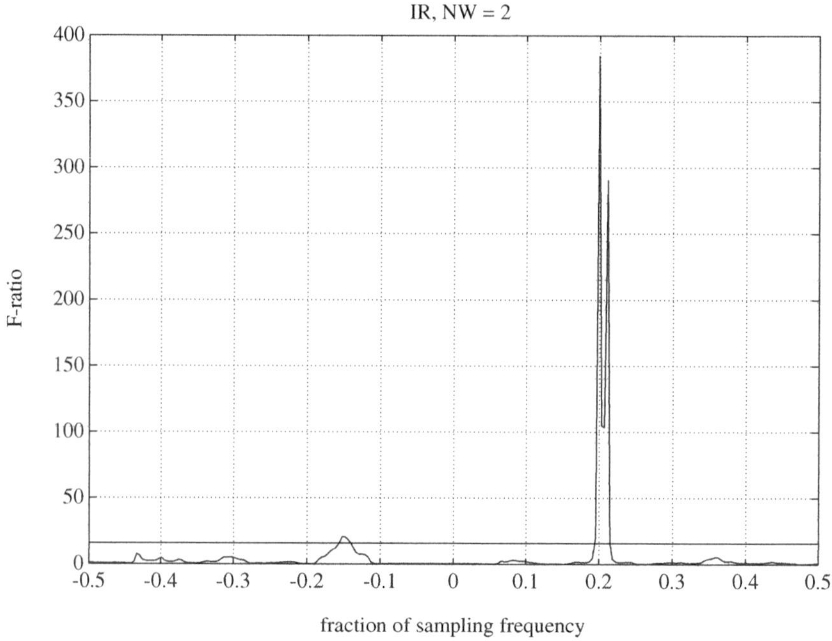

Figure 2.18 The projection of max $F(f,\Delta f)$, onto the f axis. Note the absence of spurious peaks.

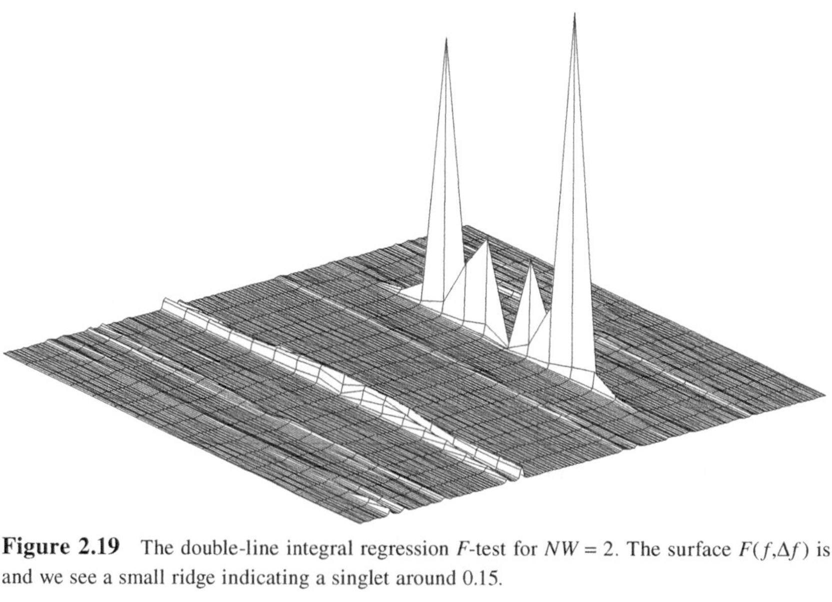

Figure 2.19 The double-line integral regression F-test for $NW = 2$. The surface $F(f,\Delta f)$ is shown and we see a small ridge indicating a singlet around 0.15.

For this residual spectrum, with no line components, the eigencoefficients $y_k(f)$ should obey the expectation

$$\mathbf{E}\{y_k(f)\} = 0$$

They should also be related to the original kth eigencoefficients $x_k(f)$ as

$$y_k(f) = x_k(f) - \sum_i \hat{\mu}(f_i)V_k(f - f_i) \tag{2.61}$$

This suggests that the data vector be modified as

$$y(n) = x(n) - \sum_i \mu_i e^{j2\pi n f_i} \tag{2.62}$$

for every n, where $\mu_i = \hat{\mu}(f_i)$. Thomson [36] gives the formula

$$S(f) = |\hat{\mu}(f_0)|^2 \delta(f - f_0) + S_r(f) \tag{2.63}$$

for reshaping the spectrum around a line component at f_0, where $S_r(f)$ is the *residual spectrum* after extracting that particular line component. In reference 3, it is also pointed out that care must be taken with this operation, so that power is conserved numerically. One cannot just stick $|\hat{\mu}(f_0)|^2$ at f_0 after estimating the residual spectrum. Since (2.63) is a power density relation, we obtain by integration

$$\hat{\sigma}^2 = \sum_i \hat{\sigma}_i^2 + \hat{\sigma}_r^2 \tag{2.64}$$

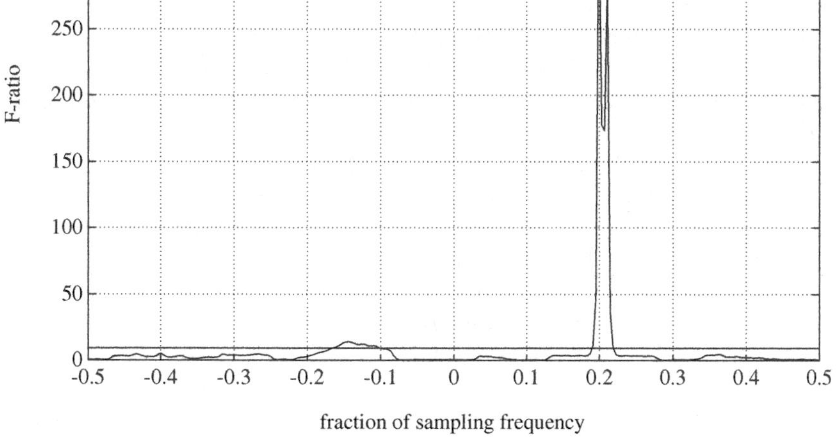

Figure 2.20 The projection of max $F(f,\Delta f)$, onto the f axis.

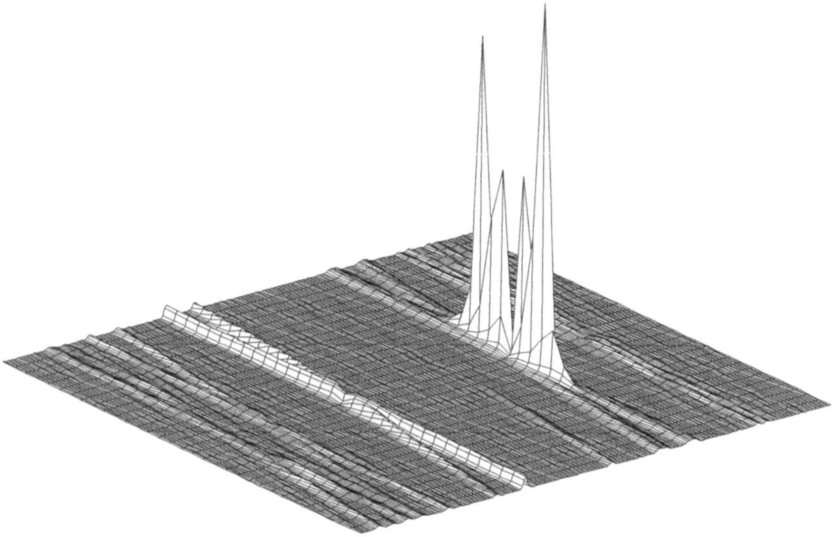

Figure 2.21 $F(f,\Delta f)$ for the double-line integral regression F-test for $NW = 4$.

where $\hat{\sigma}^2$ is our estimate of the initial process variance, $\hat{\sigma}_r^2$ is the corresponding estimate for the residual process generated by (2.62), and $\hat{\sigma}_i^2$ is the power that corresponds to the ith line component. Subtracting the residual power from the total in order to find a particular $\hat{\sigma}_i^2$ is very sensitive to roundoff errors. Hence, in implementation, it should only be used to find the total power that corresponds to all four line components (i.e., subtract initial variance from the estimated variance when all line components are extracted). The power is then allocated into four components, according to the ratios of the $|\hat{\mu}(f_i)|^2$, which exhibit better robustness properties. Since the spectrum $S(f)$ is a power density, each of those power components is assumed to be equal to the area of a rectangle of width W, the spectral window used in the estimation of the μ's. The height of this rectangle is then the proper number to assign as the line power density at the estimated line frequency. Thomson also suggests that the line component be assigned a shape, similar to the F line shape. The width of the line would then be proportional to the frequency uncertainty of the estimate. (This is not done here because the FFT mesh used is only 256 points and the F-width is approximately equal to 1/256.)

In order to perform the computations in (2.62), we require the best possible estimates of μ_i and f_i in order to minimize subtraction errors. For the univariate F function of the single-line tests, the method indicated is the *golden section search* to find the maximum of F. This is similar in spirit to the bisection method for root finding. For the double-line tests, a multivariable optimization method is required. The one used in this implementation is the *Hooke and Jeeves direct search method* (see references 11 and 14), but the specific choice of routine in both cases is a matter of personal taste.

The results of the parameter estimation of the μ_i's and f_i's are given in Tables 2.2 and 2.3. Figure 2.22 shows the residual single-line F-test when the doublet is

Table 2.2 Initial Estimates of μ_i and f_i

	First Singlet	Second Singlet	First Doublet	Second Doublet
Known frequencies	−0.15	0.1	0.2	0.21
99% level	10.92	10.92	15.98	15.98
Point Regression				
Estimated frequency	−0.1508	0.0965	0.1995	0.2100
F-ratio	66.7	17.4	8978.0	8978.0
μ	$0.0734 - j0.0654$	$0.0042 + j0.1027$	$0.2158 + j0.9335$	$0.2977 + j0.9211$
Integral Regression				
Estimated frequency	−0.1501	0.0968	0.1997	0.2099
F-ratio	49.6	5.61	1185.0	1185.0
μ	$0.0662 - j0.0726$	$0.0027 + j0.1094$	$0.2491 + j0.9404$	$0.2632 + j0.9403$

Note: A Time-Bandwidth $NW = 2$ was used. The Initial Mean and Variance Are $-0.025 + j0./024$ and 1.780, Respectively.

Table 2.3 Final Estimates of μ_i and f_i

	First Singlet	Second Singlet	First Doublet	Second Doublet
Point Regression				
Estimated frequency	−0.1499	0.1002	0.1996	0.1099
F-ratio	304.7	3040.0	10017.0	10017.0
μ	$0.0596 - j0.0829$	$0.0839 + j0.0568$	$0.2357 + j0.9362$	$0.2774 + j0.9359$
Integral Regression				
Estimated frequency	−0.1499	0.1001	0.1997	0.2098
F-ratio	201.7	915.9	1142.0	1142.0
μ	$0.0606 - j0.0839$	$0.0843 + j0.0581$	$0.2712 + j0.9437$	$0.2380 + j0.9561$

Note: The residual mean and variance are $0.0036 - j0.0007$ and 0.1821 for point regression and $0.0038 - j0.0007$ and 0.1817 for integral regression, respectively.

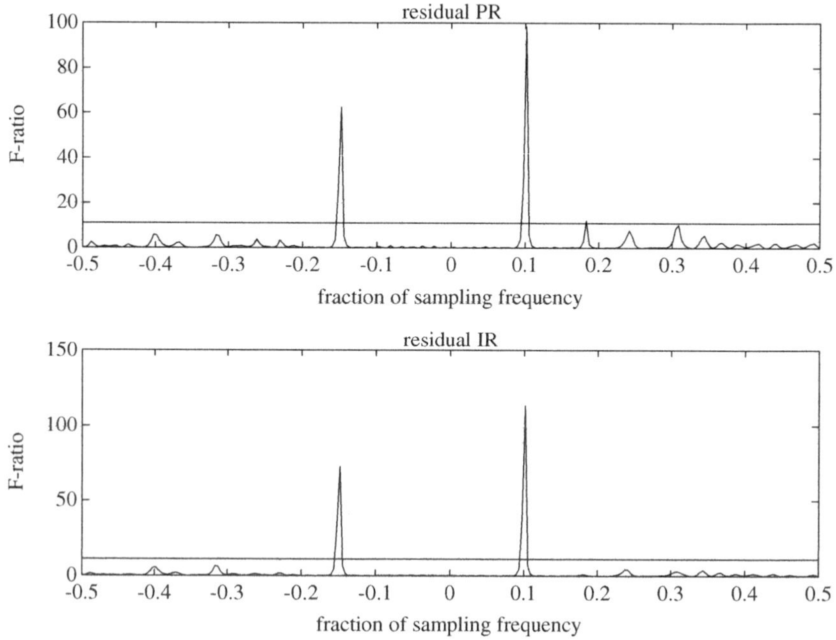

Figure 2.22 The residual single-line F-test when the initial estimate of the doublet is removed. The solid straight line indicates the 99% confidence level. Note that the spurious peak heights are reduced more in the integral regression version of the F-test.

removed. Better estimates of the singlets are then obtained. These can then be extracted from the original dataset, and the doublet can again be estimated (see Table 2.3). In fact, an iterative scheme of extracting line(s) with higher F values to improve the estimates of those with lower F values suggests itself. For more complex datasets, this may be a necessary approach.

Comparing point and integral regression parameter values, we do not see a significant difference between the final estimates of the line components. Point regression is, however, much faster in computer execution, while integral regression is more stable regarding spurious peaks.[15] The reconstructed spectrum in Fig. 2.23 is in excellent agreement with the theoretical one in Fig. 2.6. The last one gives a range of values only down to −50 dB, and we see that the only problem is some leakage of the continuous colored noise component beyond the ±0.2 frequency boundaries. This is due to the low-pass properties of the prolate windows, and Thomson [36] offers the remedy of using a composite spectrum estimator to correct for this effect. Figure 2.24 shows the result of applying this remedy; it does indeed alleviate the leakage problem, showing a more structured colored noise as well. Note

[15] It may be worthwhile to restrict integral regression, so the integral expressions are computed numerically using known eigencoefficients. This should certainly improve execution time over the computation of the double and triple sums appearing in the test.

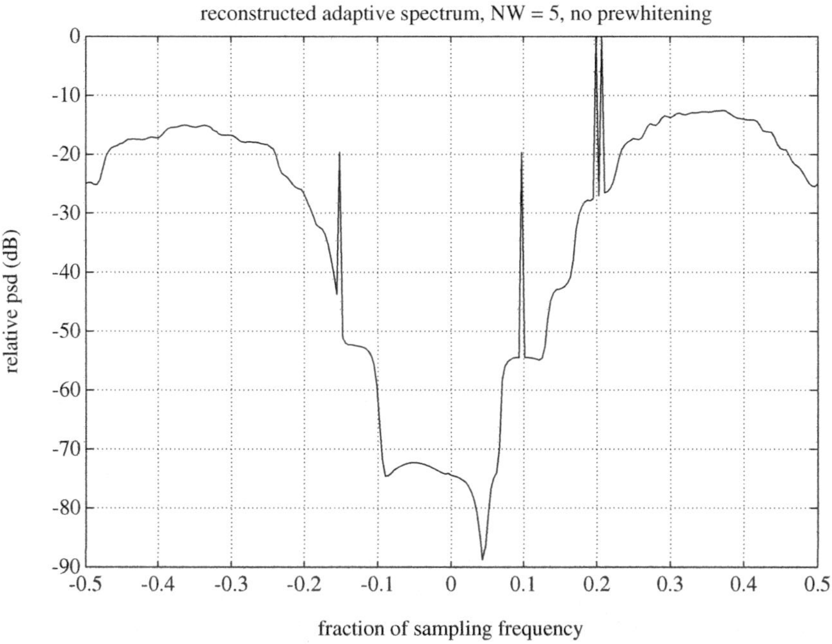

Figure 2.23 Adaptive spectrum reconstruction without prewhitening. The agreement with the theoretical analytic form is excellent except for some leakage of the colored noise beyond the ±0.2 frequency boundaries.

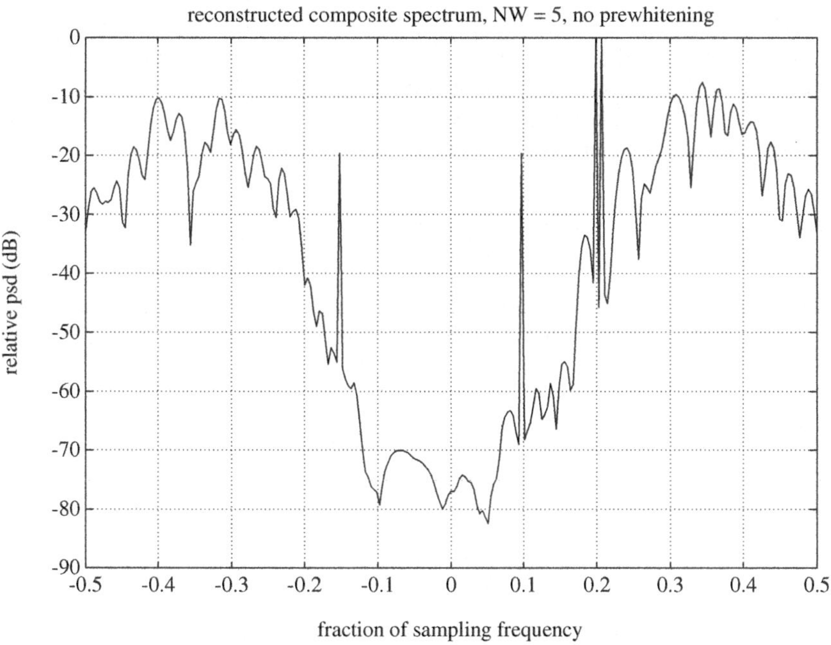

Figure 2.24 Composite spectrum reconstruction without prewhitening. The leakage problem is alleviated, and more structure is evident in the colored noise component.

that this new structure is actually more realistic to expect, since we are given only a single realization of a stochastic process.

2.7.7 Prewhitening

The importance of *prewhitening* cannot be stressed strongly enough for real-life cases. In essence, prewhitening reduces the dynamic range of the spectrum by filtering the data, so the residual spectrum is nearly flat or white. Leakage from strong components is then reduced, so the fine structure of weaker components is more likely to be resolved. Most of the theory behind spectrum estimation assumes smooth, almost white-like spectra to begin with, anyway.

In order to whiten the process $\{y(n)\}_{n=1}^{N}$, the method in vogue is to pass it through a linear prediction filter. This is equivalent to assuming an AR model for the process, and if this is true and the filter order is the same as the AR process order, then the output of the filter, $\{r(n)\}$, *the innovations process*, will be white noise. If not, the range of the spectrum will be decreased and any fine structure on the innovations process will be easier to pick out. In essence, a systems identification approach is taken. The reason that AR models are so popular is simply that they are linear and much easier to solve than MA or the full ARMA models.[16]

It should be stressed that the use of an AR filter for prewhitening does not imply an AR spectrum estimation approach. The AR filter simply performs some data preprocessing, so MTM (or any spectrum estimation technique for that matter) will operate on data having properties closer to the ones that the method or technique assumes. This makes for a better intermediate spectrum estimate. The effect of the AR filter is then removed by operating on this intermediate spectrum estimate with an "inverse filter." Note also that the prewhitened data do not have to have a purely white spectrum. It need only be relatively flat.

Therefore, assuming that

$$y(n) = -\sum_{k=1}^{p} a_k y(n-k) + r(n)$$

where the a_k's are the AR coefficients with p denoting the model order, we compute the residual spectrum $\hat{S}_r(f)$ with MTM and estimate $S_y(f)$ from the formula

$$\hat{S}_y(f) = \frac{\hat{S}_r(f)}{\left|1 + \sum_{k=1}^{p} a_k e^{-j2\pi fk}\right|^2} \tag{2.65}$$

Estimation of the AR coefficients was done with the modified covariance method, in which both the forward and backward prediction errors are minimized at the

[16] In reference 19 the authors explain that they are investigating the generalization to the full ARMA case.

same time. Marple [22] claims that this method seems to be optimum, although it does not guarantee minimum phase for the resultant filter. Marple also offers a routine that implements it (among other methods), and this is the one used here. Note that with this method, the property that the backward error coefficients are complex conjugates of the forward error coefficients holds. The identification of the latter with the AR parameters helps us compute the residuals without assuming zero initial conditions, which introduce transient errors to the process as

$$r(n) = \begin{cases} y(n) + \sum_{k=1}^{p} a_k y(n-k) & \text{for } n = p+1, \ldots, N \\ y(n) + \sum_{k=1}^{p} a_k^* y(n+k) & \text{for } n = 1, \ldots, p \end{cases} \tag{2.66}$$

Observe also that in this way, we do not reduce the sample size of the given time series.

A straightforward implementation of the above equation leads to Figs. 2.25 and 2.26 for $NW = 5$ and an AR(4) model, using adaptive reconstruction and composite reconstruction, respectively. Prewhitening brings more structure to the colored noise, which is what we would realistically expect. Note that in the case of the composite spectrum, it is evident that the extraction of the singlets is not total, since we see some increase in the residual spectrum at those frequencies. Prewhitening brings this detail out. Note also that if prewhitening is performed before the line components are estimated and we then try to estimate them from the prewhitened data, some small

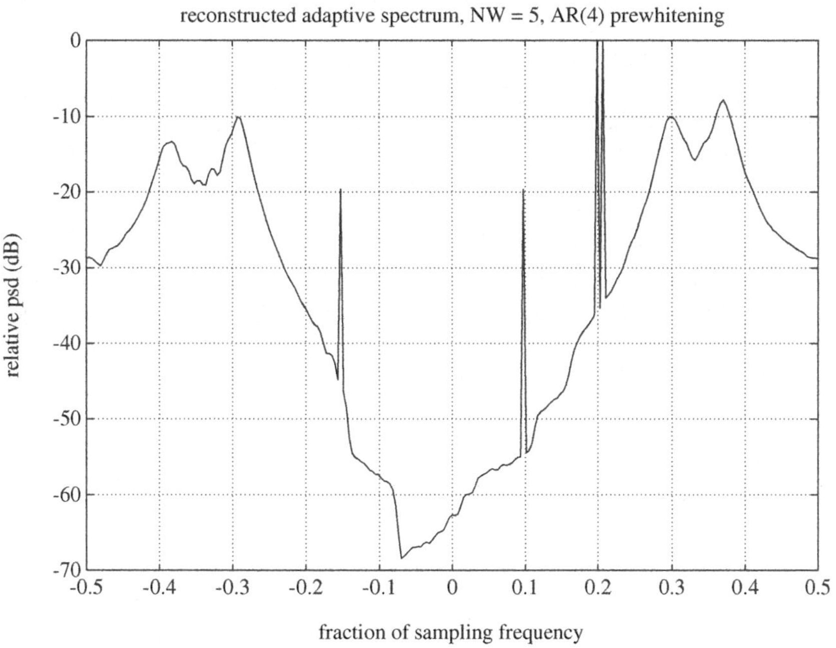

Figure 2.25 The adaptive reconstructed spectrum with prewhitening.

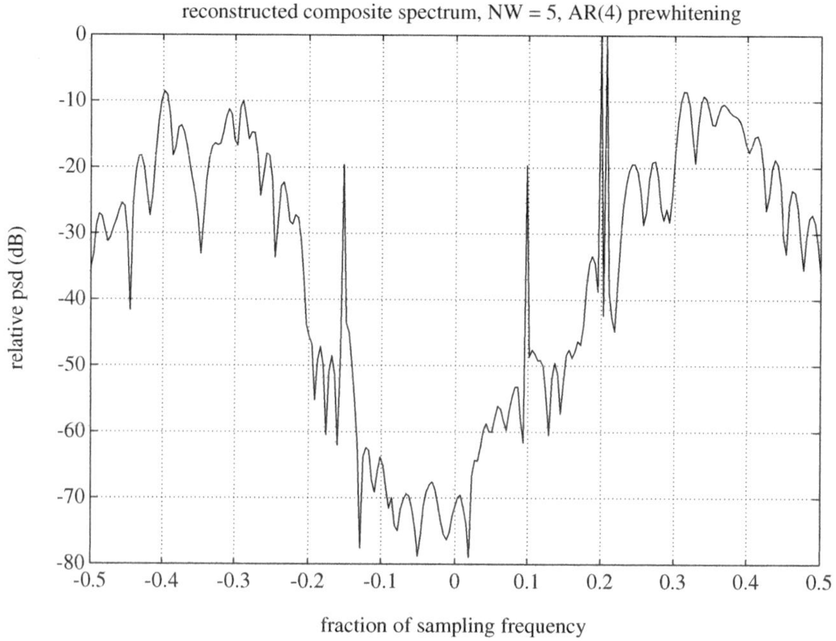

Figure 2.26 The composite reconstructed spectrum with prewhitening.

improvement is indeed seen with the weaker, singlet components. However, the stronger doublet is removed by the AR filter. This indicates that the approach described herein be used only for estimating weak power-line components.

Comparing Figs. 2.23–2.26 with the ideal spectrum in Fig. 2.6, it appears that Fig. 2.23 is the best match, indicating that prewhitening does not seem to improve things. This was not the case with other datasets and spectra that were tried. Prewhitening always led to an improved spectrum estimate. The discrepancy observed with Marple's dataset is probably coincidental. It so happens that the smoothing that the prolate windows perform give a final spectrum estimate that is very close to the ideal one. Normally, more than one snapshot is required for such an event to occur. Looking into the classical spectrum estimation results in Fig. 2.7, we do see some similarities in the colored noise component, particularly with the Hamming spectrum. Clearly, the prewhitening gives a more accurate picture of the single snapshot spectrum estimate, but, coincidentally in this case, the non-prewhitened spectrum estimate is closer to the ideal one.

In references 35 and 19, the case of robust prewhitening is worked out for a simple additive outlier model, for real data. It is assumed that the observed data $\{y_n\}_{n=1}^{N}$ consist of the process of interest $\{x_n\}_{n=1}^{N}$, plus occasional outliers $\{e_n\}_{n=1}^{N}$

$$y_n = x_n + e_n$$

and an iterative scheme is developed. A further refinement is given in reference 23. It would be informative to extend this technique to complex data, making prewhitening more robust; however, this is not done here.

2.7.8 Multiple Snapshots

In a practical situation, we always have more than one snapshot of data available. Since the underlying process is stochastic in nature, it makes sense to combine all the available snapshots in estimating the spectrum into a single overall result. Note that simply averaging the individual adaptive or composite spectra for each snapshot is a nonoptimum way of doing so. The proper generalization is described in what follows.

For a single data vector $\{x_n\}_1^N$, the K eigencoefficients $x_k(f)$ are computed, and from them the individual eigenspectra $\hat{S}_k(f) = |x_k(f)|^2$ are estimated. Generalizing to N_s data vectors (snapshots) $\{x(n, n_s)\}$, $n = 1, \ldots, N$ and $n_s = 1, \ldots, N_s$, the average eigenspectra would be

$$\hat{S}_k(f) = \frac{1}{N_s} \sum_{n_s=1}^{N_s} |x_k(f|n_s)|^2$$

It is these values that are used in the adaptive and composite spectrum estimation procedures.

The F-tests are also slightly modified. In the following subsection, the extension to multiple-snapshot point-regression F-tests is made. The integral regression case follows along similar lines.

2.7.9 Multiple-Snapshot, Single-Line, Point-Regression *F*-Tests

For the single-line, point-regression case, the criterion is the same. Given a line component at a frequency f_0, the expected value of the eigencoefficients is

$$\mathbf{E}\{x_k(f|n_s)\} = \mu V_k(f - f_0)$$

where now all N_s snapshots are included. The residual error that has to be minimized is $\|\mathbf{x} - \mathbf{U}\mu\|^2$, where

$$\mathbf{x} = \begin{bmatrix} x_0(f|1) \\ \vdots \\ x_{K-1}(f|1) \\ \vdots \\ x_0(f|N_s) \\ \vdots \\ x_{K-1}(f|N_s) \end{bmatrix}, \qquad \mu = \mu, \qquad \mathbf{V} = \begin{bmatrix} V_0(f - f_0) \\ V_1(f - f_0) \\ \vdots \\ V_{K-1}(f - f_0) \end{bmatrix}, \qquad \mathbf{U} = \begin{bmatrix} \mathbf{V} \\ \mathbf{V} \\ \vdots \\ \mathbf{V} \end{bmatrix}$$

The data vector $\mathbf{x}$ is now a $KN_s \times 1$ complex vector, μ is again a 1×1 complex scalar, $\mathbf{V}$ is the same $K \times 1$ complex vector, but $\mathbf{U}$ is now introduced, containing N_s identical $\mathbf{V}$ vectors.

The residual error at f can be written in component form as the squared term

$$e^2(\mu, f) = \sum_{n_s=0}^{N_s} \sum_{k=0}^{K-1} |x_k(f|n_s) - \mu(f)V_k(0)|^2$$

and the least-squares solution is

$$\hat{\mu} = \left(\mathbf{U}^H\mathbf{U}\right)^{-1}\mathbf{U}^H\mathbf{x} = \hat{\mu}(f) = \frac{\sum_{n_s=1}^{N_s}\sum_{k=0}^{K-1} V_k^*(0)x_k(f|n_s)}{N_s\sum_{k=0}^{K-1}|V_k(0)|^2}$$

This expression can also be written in the compact form

$$\hat{\mu}(f) = \sum_{n=0}^{N-1} h_n(N, W)\overline{x}(n)e^{-j2\pi fn}$$

The nth harmonic data window is defined by

$$h_n(N, W) = \frac{\sum_{k=0}^{K-1} V_k^*(0)\upsilon_n^{(k)}(N, W)}{\sum_{k=0}^{K-1}|V_k(0)|^2}$$

which the same as in the single snapshot case; see (2.40). But we now have

$$\overline{x}(n) = \frac{1}{N_s}\sum_{n_s=1}^{N_s} x(n|n_s)$$

as the coherent average. Correspondingly, the F-ratio is defined by

$$F(f) = \frac{\frac{1}{\nu_1}\|\mathbf{U}\mu\|^2}{\frac{1}{\nu_2}\|\mathbf{x} - (\mathbf{U})\mu\|^2}$$

or

$$F(f) = \frac{\frac{1}{\nu}|\hat{\mu}(f)|^2 N_s \sum_{k=0}^{K-1}|V_k(0)|^2}{\frac{1}{2K-\nu}\sum_{n_s=1}^{N_s}\sum_{k=0}^{K-1}|x_k(f|n_s) - \mu(f)V_k(0)|^2}$$

where ν is again equal to two degrees of freedom. The total number of degrees of freedom, ν_{tot}, is set here equal to the old value $2K$. This is a worst-case scenario. We do have KN_s complex data points now available to draw information from. However, using the value $\nu_{tot} = 2KN_s$ for multiple snapshot F-tests, with experimental data (see later sections), we obtain preposterously large F values, orders of magnitude larger than the 99% confidence level. There is no doubt that we do have more degrees of freedom but the data snapshots may not be independent enough, one from the other, in order to assign $\nu_{tot} = 2KN_s$. The appropriate value for ν_{tot} is probably somewhere between $2K < \nu_{tot} < 2KN_s$. In reference 38, p. 565, Thomson points out that the effective number of degrees of freedom is given by a more complicated expression and is indeed less than KN_s (development of this formula is not done here).

Actually, the proper number of degrees of freedom is only needed to decide on the proper significance level. Using the test as is, we can check the relative heights of the various F peaks. Since we already know the number and approximate location, we simply pick the largest peaks in that region. In the real data cases examined in Sections 2.8 and 2.9, it is seen that excellent estimates are obtained with this approach, both for the line frequencies and their complex amplitudes. However, in order to test for the existence of other possible line components, this matter should be resolved.

2.7.10 Multiple-Snapshot, Double-Line Point-Regression *F*-Tests

Similar changes occur for the double-line tests. We have a similar criterion as before, namely,

$$\mathbf{E}\{x_k(f|n_s)\} = \mu_1 V_k(f-f_1) + \mu_2 V_k(f-f_2)$$

where now all N_s snapshots are included. The residual error to be minimized is $\|\mathbf{x} - \mathbf{U}\boldsymbol{\mu}\|^2$, where

$$\mathbf{x} = \begin{bmatrix} x_0(f|1) \\ \vdots \\ x_{K-1}(f|1) \\ \vdots \\ x_0(f|N_s) \\ \vdots \\ x_{K-1}(f|N_s) \end{bmatrix}$$

$$\boldsymbol{\mu} = \begin{bmatrix} \mu_1 \\ \mu_2 \end{bmatrix}$$

$$\mathbf{V} = \begin{bmatrix} V_0(f-f_1) & V_0(f-f_2) \\ V_1(f-f_1) & V_0(f-f_2) \\ \vdots & \vdots \\ V_{K-1}(f-f_1) & V_{K-1}(f-f_2) \end{bmatrix}$$

$$\mathbf{U} = \begin{bmatrix} \mathbf{V} \\ \mathbf{V} \\ \vdots \\ \mathbf{V} \end{bmatrix}$$

If again, as in the single-snapshot case

$$\mathbf{V}^H\mathbf{V} = \begin{bmatrix} d_1 & c \\ c^* & d_2 \end{bmatrix}$$

then the least-squares solution is

$$\boldsymbol{\mu} = \left(\mathbf{U}^H\mathbf{U}\right)^{-1}\mathbf{U}^H\mathbf{x} = \frac{1}{d_1 d_2 - |c|^2}\begin{bmatrix} d_2 & -c \\ -c^* & d_1 \end{bmatrix}\begin{bmatrix} c_1 \\ c_2 \end{bmatrix}$$

where we now have

$$d_1 = \sum_{k=0}^{K-1} |V_k(f - f_1)|^2, \qquad d_2 = \sum_{k=0}^{K-1} |V_k(f - f_2)|^2$$

$$c = \sum_{k=0}^{K-1} V_k^*(f - f_1)V_k(f - f_2)$$

$$c_1 = \frac{1}{N_s}\sum_{n_s=1}^{N_s}\sum_{k=0}^{K-1} V_k^*(f - f_1)x_k(f|n_s)$$

$$c_2 = \frac{1}{N_s}\sum_{n_s=1}^{N_s}\sum_{k=0}^{K-1} V_k^*(f - f_2)x_k(f|n_s)$$

The F-test, for testing the hypothesis that at a given frequency we need two lines to explain the data, becomes

$$F(f; f_1, f_2) = \frac{1/(\nu_1 - \nu_2)}{1/(2K - \nu_1)}\left[\frac{\|\mathbf{U}\boldsymbol{\mu}\|_{\text{double}}^2 - \|\mathbf{U}\boldsymbol{\mu}\|_{\text{single}}^2}{\|\mathbf{x} - \mathbf{U}\boldsymbol{\mu}\|_{\text{double}}^2}\right]$$

The single-line model parameter μ is now given by

$$\mu_{\text{single}} = c_1/d_1$$

and

$$\|\mathbf{U}\boldsymbol{\mu}\|_{\text{single}}^2 = N_s\,|c_1|^2/d_1$$

while

$$\|\mathbf{U}\boldsymbol{\mu}\|_{\text{single}}^2 = N_s\sum_{k=0}^{K-1} |V_k(0)\mu_1 + V_k(f - f_2)\mu_2|^2$$

Similar observations, as given previously, hold for ν_{tot}, the effective number of degrees of freedom.

2.8 EXPERIMENTAL DATA DESCRIPTION FOR A LOW-ANGLE TRACKING RADAR STUDY

In this section, we apply the multi-taper method to real-life sampled-aperture data corresponding to a low-angle tracking radar environment. The experimental data collection was performed at a site on the mouth of Dorcas Bay, which opens onto the eastern end of Lake Huron on the west coast of the Bruce Peninsula, close to Tobermory, Ontario. This particular location was chosen because of the high sea states normally encountered there, which are caused by a combination of westerly winds, the shallow water offshore,[17] and the long fetch across Lake Huron. The transmitter was located at a distance $L = 4.61$ km from the receiver, both being within 10 m of the water's edge. Figure 2.27 shows the path profile. The receiver was secured at the top of a tower and its center was at a height h_R above the water's edge, while that of the transmitter was at some adjustable value denoted by h_T.

The transmitter consists of two antenna horns, one above the other, the top one being set for horizontal (H) and the bottom one for vertical (V) polarizations, respectively. Each horn was also capable of transmitting at different frequencies in order to implement frequency agility. For the latter, two different frequency signals were transmitted simultaneously. One of them was always at the fixed frequency of 10.2 GHz, while the other frequency was varied from 8.02 to 12.34 GHz in 30-MHz steps, according to the formula: $8.02 + 0.03p$ GHz. The picket number p controlled the frequency step size in the agile channel.

The available options for the transmitted signal were chosen as follows:

- When a horizontally polarized signal was desired, the top H horn would transmit the mixture of the fixed and agile frequency signals, while the bottom V horn was idle.
- For an H and V dual-polarized signal, the H horn would transmit the agile-frequency signal and the V horn would transmit the fixed-frequency signal.
- Finally, for a vertically polarized signal, the top H horn remained idle while the bottom V horn transmitted the mixture of fixed- and agile-frequency signals.

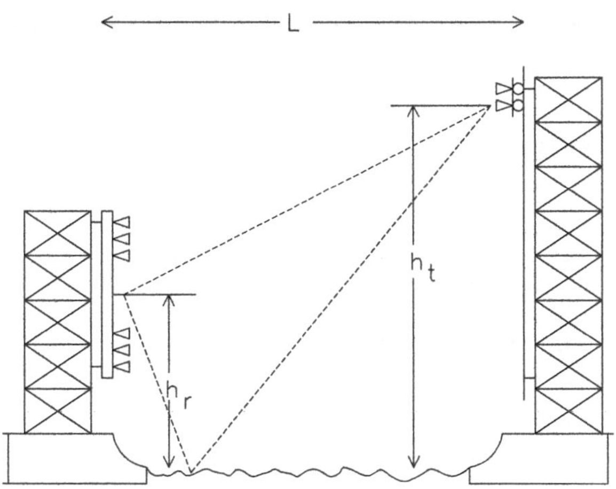

Figure 2.27 The path profile: transmitter is located at the right, and 32-element receiver array is located at the left.

[17] The greatest depth along the transmission path was 12 m.

As described in Chapter 1, the microwave source was produced by phase-locking to a highly stable 5-MHz crystal oscillator, operating at a constant temperature. A traveling-wave tube amplifier (TWTA) with a maximum gain of 40dB was then used to amplify the source signal before feeding it to the transmitter horns. System control at the transmitter was achieved with a PC-XT microcomputer that adjusted transmission frequency, TWTA gain, and carriage height.

As mentioned in the Introduction, the receiver was a 32-element linear array, forming the main part of MARS (Table 2.4). More specifically, MARS is a coherent, sampled aperture, linear receiver array (vertically oriented), signal-processing system, composed of 32 pairs of horizontally polarized standard gain horn antennas. Each pair of horns corresponds to a single-array element and offers the opportunity to always implement frequency agility and when desired, by choosing one of the options for the polarization of the transmitted signal, polarization diversity, HH, or VH.

The RF and IF local oscillators and the RF testing signal generator were shared by all the receiver elements. Since we have coherent reception, each receiver horn is followed by a quadrature demodulator block, so that phase information can be extracted.

System control at the receiver was provided with a microcomputer that offered a completely automated data collection environment. For each trial, only the parameters defining the desired physical conditions would need to be fed into the computer. The computer accordingly adjusted the frequency of the RF oscillator for the agile channels, the IF amplifier gains, the bandwidth of the low-pass filters in the I and Q channels, and the sampling rate of the samplers. It also sent commands via a radio link to the transmitter computer that controlled the transmission operation.

Table 2.4 Specifications of MARS, Used in the Lake Huron Experiments

Transmitter
- 100-mW CW into one or two 22-dB horns
- Simultaneous dual frequencies
- Option of dual polarization (H) or (V)
- Adjustable transmitter height h_T

Antenna Array
- 32-element linear array vertically oriented
- Array aperture 1.82 m, interelement spacing 5.715 cm
- Array machined to ±0.1-mm tolerance
- Multifrequency capability 8.02–12.34 GHz with 30-MHz steps

Receiver Elements
- Two receiver channels per array element
- 10-dB horizontally polarized horns as array elements
- ~25-dB nominal cross polarization rejection ratio
- Coherent demodulation with nominal frequency stability to 10^{-12} over a few seconds
- 0.1-Hz Doppler resolution
- 1-Hz to 2-kHz sampling rate capability

Most of the data were sampled with a 62.5-Hz sampling rate, corresponding to 8 sample points per cycle at the baseband frequency. They were then digitized by A/D converters with 12-bit precision and stored onto hard disk for off-line processing.

The data were divided into three groups with respect to polarization: like-polarized, dual-polarized, and cross-polarized. The first four characters in their name referred to the collection date—for example, *nov3*. The next two characters define the group to which they belong; *dh* for the like-polarized group where both transmission and reception are H-polarized; *dd* for the dual-polarized group where both polarizations are transmitted; *dv* for the cross-polarized group where only a cross polarized signal is received; and *cff* for the far-field, aperture calibration datasets.

The numeral following the polarization designation is simply a dataset index providing order among similar datasets collected within the same day. It was followed by the data subset number. As an example, *nov3:dd1.dat;2* would stand for the second subset of a dual-polarized dataset, collected on November 3, being the first dataset in this group collected on that particular day. We normally have 127 snapshots per data subset (offering us about 2-s worth of data) and 16 subsets per dataset. Table 2.5 presents some sample datasets collected with MARS.

The data were carefully calibrated, both in phase and quadrature (IQ calibration) using known injected tone signals, and in aperture (aperture calibration) using a far-field technique (details are given in reference 8). In the following section, we focus on some HH polarized data at different frequencies and grazing angles. The particular datasets used are shown in Table 2.5. Note that a spherical earth model [16, p. 4] where normal propagation conditions are assumed—earth radius is 4/3 the actual value—is used to calculate the expected angular separation between the direct and specular components. This quantity provides a better comparison measure between expected and estimated results, since it is not affected by the tilt of the array antenna occurring from wind action on the array tower.

2.9 ANGLE-OF-ARRIVAL (AOA) ESTIMATION

Angle-of-arrival estimation on the MARS database was performed using both the multi-taper method (MTM) and the maximum likelihood (ML) method, on a

Table 2.5 A Sample of Individual Experimental Datasets Used in the MTM Implementation[a]

Dataset	Frequency (GHz)	Separation (BW)	h_r (m)	h_t (m)
1. *nov4:dh9.dat;1*	8.05	0.320	8.64	15.53
2. *nov3:dh4.dat;3*	8.62	0.218	8.67	9.59
3. *nov3:dh6.dat;7*	9.76	0.247	8.67	9.59
4. *nov3:dd1.dat;2*	10.12	0.468	8.67	18.06
5. *nov3:dh4.dat;16*	12.34	0.312	8.67	9.59

[a] The distance between transmitter and receiver is 4.61 km in all cases.

snapshot-by-shapshot basis. ML was chosen because it provides the yardstick against which all modern parametric methods are usually compared. To single out the effect, if any, of including diffuse multipath in the AOA estimation process, which MTM implicitly does, the only *a priori* knowledge that we assume for ML is that we have two incoming plane waves in a white noise background. The noise covariance matrix is then diagonal, and the maximization of the logarithm of the likelihood function leads to the minimization

$$\min_{\mathbf{\Phi}} \|\mathbf{x}-\mathbf{s}\|^2 \tag{2.67}$$

This is a nonlinear least-squares problem, where $\mathbf{\Phi}$ is the parameter vector, $\mathbf{x}$ is the data vector, and $\mathbf{s}$ is the signal model. Let us express the signal vector as the matrix product

$$\mathbf{s} = \mathbf{A}\mathbf{a} \tag{2.68}$$

where $\mathbf{A}$ is a function of the parameters that enter nonlinearly into the model, while $\mathbf{a}$ is a function of the ones that do so linearly. Then, (2.67) becomes

$$\min_{\mathbf{\Phi}} \|\mathbf{x}-\mathbf{A}\mathbf{a}\|^2 \tag{2.69}$$

The separation of $\mathbf{\Phi}$ into a linear and a nonlinear subset helps us break the nonlinear least-squares estimation problem of (2.67) into a linear least-squares problem and a smaller nonlinear one that is hopefully easier to solve. This is done as follows. At the minimum, when $\mathbf{a} = \hat{\mathbf{a}}$, the residual vector $\mathbf{x} - \mathbf{A}\hat{\mathbf{a}}$ is orthogonal to the product term $\mathbf{A}\mathbf{a}$. Therefore,

$$(\mathbf{A}\mathbf{a})^H(\mathbf{x}-\mathbf{A}\hat{\mathbf{a}}) = 0 \tag{2.70}$$

and

$$\hat{\mathbf{a}} = \mathbf{A}^+\mathbf{x} = \left(\mathbf{A}^H\mathbf{A}\right)^{-1}\mathbf{A}^H\mathbf{x} \tag{2.71}$$

Also, from (2.69) we have

$$\begin{aligned}\|\mathbf{x}-\mathbf{A}\mathbf{a}\|^2 &= (\mathbf{x}-\mathbf{A}\mathbf{a})^H(\mathbf{x}-\mathbf{A}\mathbf{a}) = \left[\mathbf{x}^H-(\mathbf{A}\mathbf{a})^H\right](\mathbf{x}-\mathbf{A}\mathbf{a})\\ &= \mathbf{x}^H\mathbf{x}-\mathbf{x}^H\mathbf{A}\mathbf{a}-(\mathbf{A}\mathbf{a})^H(\mathbf{x}-\mathbf{A}\mathbf{a})\end{aligned}$$

At the minimum, $\mathbf{a} = \hat{\mathbf{a}}$ and from (2.70) the last part is zero. Therefore, the condition (2.69) takes the new form

$$\max_{\mathbf{A}} \mathbf{x}^H\mathbf{A}\mathbf{A}^+\mathbf{x} \tag{2.72}$$

where the maximization is done over the nonlinear parameter subset. In problems of this type, we first solve (2.72) and then (2.71), which is now a simple linear problem.

For the case of two plane waves impinging on an M-element array, the signal model at the mth element, referred to as the array center, is

$$s(m) = a_1 e^{j[m-((M+1)/2)]\phi_1} + a_2 e^{j[m-((M+1)/2)]\phi_2}, \qquad m = 1, \ldots, M \tag{2.73}$$

where ϕ, the electrical phase angle or spatial wavenumber, is defined as [10]

$$\phi = \left(\frac{2\pi d}{\lambda}\right)\sin\theta$$

with d denoting the array interelement spacing, λ the wavelength of the signal, and θ the physical direction of arrival of the incoming plane wave.

It can be seen that the signal is linear in the complex amplitudes a_1 and a_2 and nonlinear in the phase angles ϕ_1 and ϕ_2. To change the formulation to vector notation, define the *direction of arrival* (DOA) *vector*

$$\mathbf{d}(\phi) = \left[e^{j[1-(M+1)/2]\phi}, e^{j[2-(M+1)/2]\phi} \dots e^{j[M-(M+1)/2]\phi}\right]^T \tag{2.74}$$

The *direction matrix* for this case, where the number of plane waves is $K = 2$, becomes

$$\mathbf{A} = [\mathbf{d}(\phi_1), \mathbf{d}(\phi_2)]$$

and the *signal amplitude vector* is

$$\mathbf{a} = [a_1, a_2]^T$$

The signal vector is then, as before, expressed as $\mathbf{s} = \mathbf{Aa}$, and we end up having to solve (2.72) and (2.71). Some further simplification can be done by defining the complex scalars

$$D_k = \mathbf{d}^H(\phi_k)\mathbf{x} \quad \text{and} \quad p = \mathbf{d}^H(\phi_1)\mathbf{d}(\phi_2)$$

where $k = 1, 2$. We then have

$$\mathbf{A}^H\mathbf{x} = [D_1 \quad D_2]^T$$

and

$$\left(\mathbf{A}^H\mathbf{A}\right)^{-1} = \begin{pmatrix} M & p \\ p^* & M \end{pmatrix}^{-1} = \left(M^2 - pp^*\right)^{-1}\begin{pmatrix} M & -p \\ -p^* & M \end{pmatrix}$$

The nonlinear problem is now transformed into a maximization problem, where we have to maximize the real scalar function

$$f(\phi_1, \phi_2) = \frac{|D_1|^2 + |D_2|^2 - 2\,\mathbf{Re}\left[pD_1^*D_2/M\right]}{M^2 - |p|^2} \tag{2.75}$$

The desired AOAs are the ϕ_1, ϕ_2 values that maximize $f(\phi_1, \phi_2)$. The function $f(\phi_1,\phi_2)$ is a particularly tough one to handle numerically, since in the neighborhood of $\Delta f = \phi_2 - \phi_1 \sim 0$, the denominator of (2.75) becomes close to zero. As $\Delta\phi \to 0$, the function takes the indeterminate form 0/0. A double application of L'Hôpital's rule gives us then the finite limit [17]

$$\lim_{\phi_2 \to \phi_1} f(\phi_1, \phi_2) = \frac{12|D_1'|^2 + \left(M^2 - 1\right)|D_1|^2}{M^2\left(M^2 - 1\right)}$$

where the prime applied to D_1 indicates differentiation of D_1 with respect to ϕ_1.

Looking at the behavior of the objective function for a given ϕ_1 and a particular data snapshot, despite double precision arithmetic, we see the onset of oscillations (Fig. 2.28) as ϕ_2 approaches ϕ_1 from both left and right. An optimization package, given this behavior, could converge to a nonoptimum pair of values. An asymptotic expansion in the troublesome region suggests itself. The one used here is the Taylor's expansion of $f(\phi_1, \phi_2)$ up to the linear order term:

$$f(\phi_1, \phi_2) \sim f(\phi_1, \phi_2)|_{\phi_2=\phi_1} + (\phi_2 - \phi_1)\frac{\partial f}{\phi_2}\bigg|_{\phi_2=\phi_1}$$

The linear term is another 0/0 indeterminate form requiring the application of L'Hôpital's rule five times to obtain

$$\frac{\partial f}{\phi_2}\bigg|_{\phi_2\to\phi_1} = \frac{2\sum_k\sum_l x_k^* x_l e^{j(k-l)\phi_1} j\left\{k-l\left[3\left(k-\frac{M+1}{2}\right)\left(l-\frac{M+1}{2}\right)+\frac{M^2-1}{4}\right]\right\}}{M^2(M^2-1)}$$

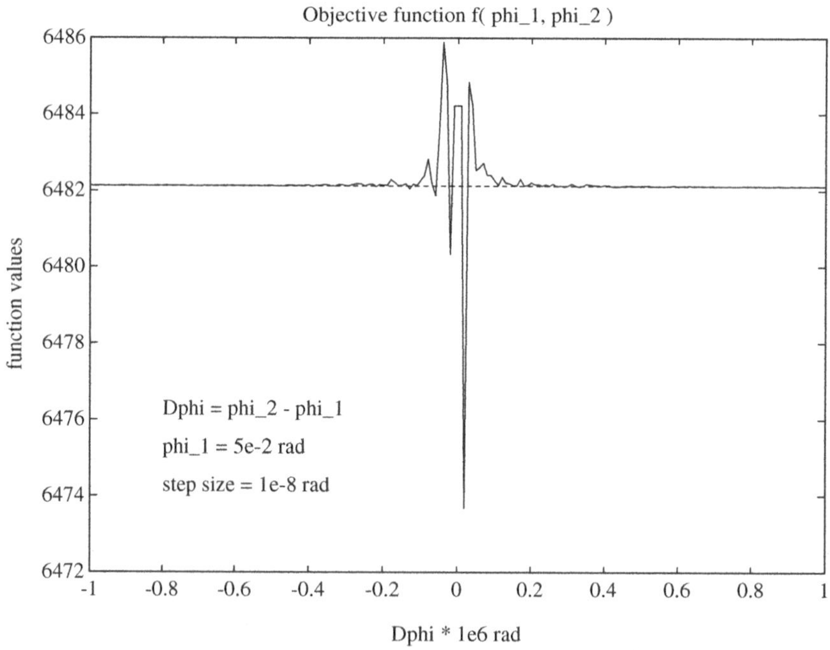

Figure 2.28 Behavior of the objective function (solid line) and its linear asymptotic expansion (dashed line) as $\Delta\phi \to 0$. Note that the function value at $\Delta\phi = 0$ is removed in the unmodified form of the function.

which is an antisymmetric expression, giving a real value. As we can see in Fig. 2.28, the numerical problem has indeed been resolved.

Given the pair (ϕ_1, ϕ_2) that maximizes the objective function f, we can now estimate the signal amplitude vector as

$$\hat{\mathbf{a}} = \frac{\begin{pmatrix} M & -p \\ -p^* & M \end{pmatrix}\begin{pmatrix} D_1 \\ D_2 \end{pmatrix}}{M^{-2}\,|p|^2}$$

Unfortunately, the method becomes more complex when the number of assumed plane waves is larger than two. Again, the function $f(\phi_1,\phi_2)$ is almost flat in the neighborhood of the maximum when ϕ_1 is close to ϕ_2. In many cases also, where the model mismatch is particularly strong (strongly colored noise), the optimization algorithm converges to a solution where $|\phi_1 - \phi_2|$ is less than 0.001 of a beamwidth (such solutions are rejected).

Applying both MTM and ML to the datasets in Table 2.5, we obtain the results shown in Figs. 2.29–2.38. The signal ratio $\rho = a_2/a_1$ (magnitude and phase) is also included. This is related to the reflection coefficient R of the scattering surface. Ideally, for a plane water surface at 10 GHz, for HH polarization and low-grazing angles, the reflection coefficient R should have a magnitude approximately equal to 0.9 and a phase of approximately 180°. For a rough surface, the mean magnitude is modified by the exponential [2]

$$\exp\left[-\left(\frac{4\pi\sigma_h \sin\psi}{\lambda}\right)^2\right]$$

where ψ is the grazing angle and σ_h is the rough surface height variance. The magnitude of ρ can be modeled by the modified R above; and, possibly, the effects of the earth's curvature can be included by using the divergence factor.

The phase of ρ, on the other hand, includes both the phase of R and the path length difference between the direct and specular paths, which may result in a shift in the observed phase, away from 180°. The quantity ρ, the specular model reflection coefficient, provides a way to check the data calibration. Occasionally, some far-field calibration data are so erroneous as to give $|\rho| > 1$. Because this would imply that the specular component is more powerful than the direct signal, data calibrated with these tables were rejected.

Examining the estimated AOAs, we can make the following observations for each dataset individually:

Dataset 1. No significant difference is observed between MTM and ML estimates. The direct component, being more powerful than the specular component, displays a smaller variance than the specular one. A rather large bias is observed between the expected and estimated AOA separation. Comparing with results reported in reference 21, where this same dataset was used with a more refined ML approach, we see that the bias appears in the estimation of the specular component (at 8.05 GHz, one BW corresponds to 1.17°). Note

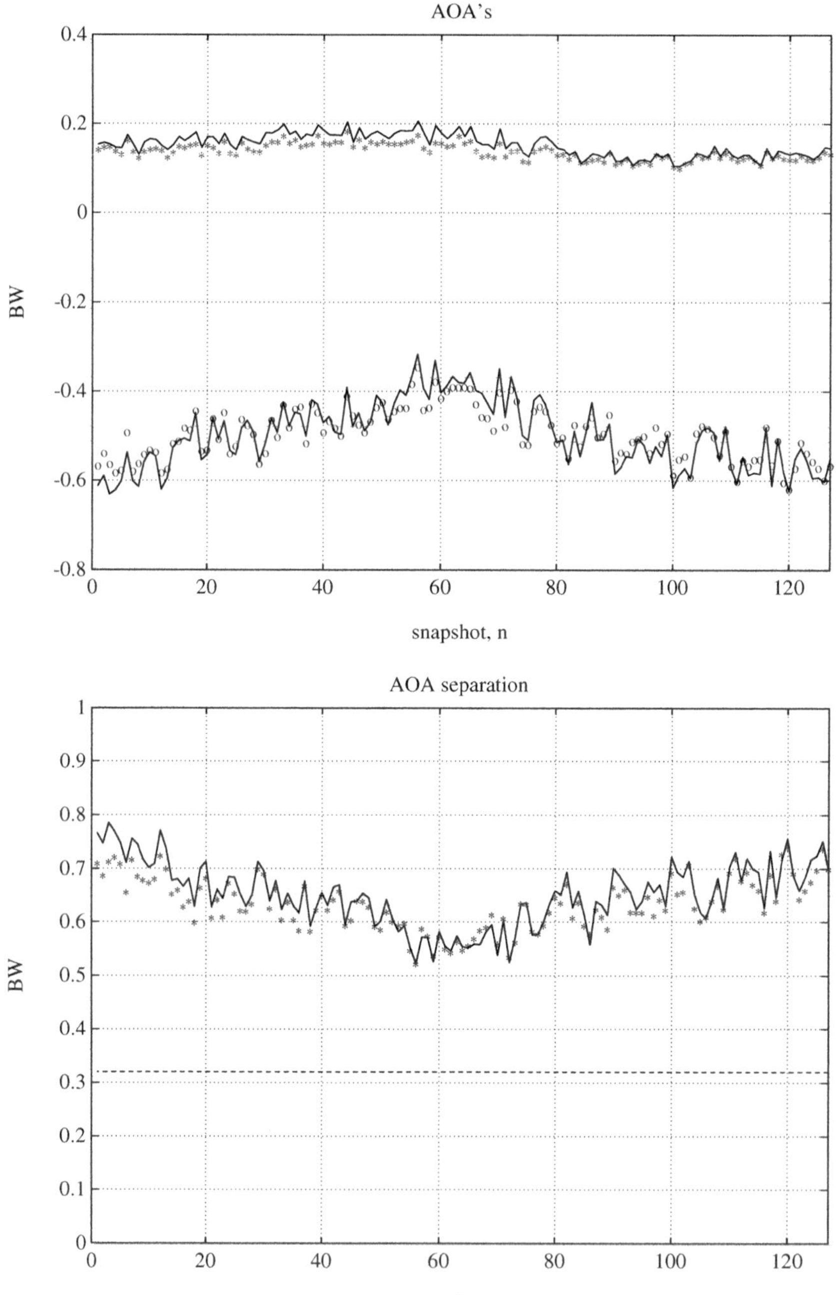

Figure 2.29 Dataset 1, *nov4:dh9,dat;1*, $f = 8.05\,\text{GHz}$, far field calibrated by *nov2:cffl.dat;1*. Solid lines always correspond to MTM estimates. In the top graph the direct component is above the specular, and * and ° correspond to the ML direct and specular components, respectively. The second graph shows the estimated angular separation between direct and specular components compared with the expected one (dashed line).

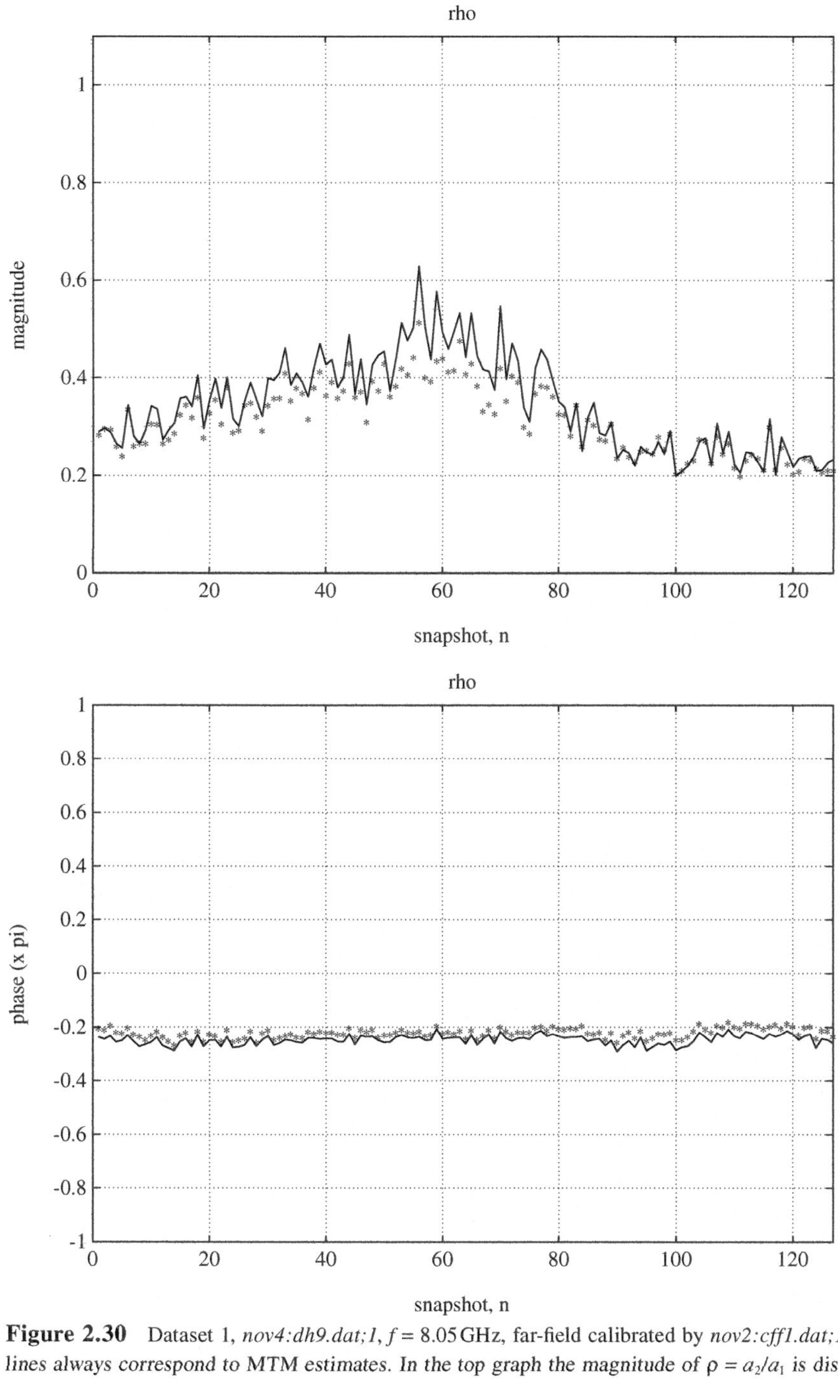

Figure 2.30 Dataset 1, *nov4:dh9.dat;1*, $f = 8.05\,\text{GHz}$, far-field calibrated by *nov2:cff1.dat;1*. Solid lines always correspond to MTM estimates. In the top graph the magnitude of $\rho = a_2/a_1$ is displayed, while in the bottom one it is the phase. The * in both cases corresponds to the ML estimates.

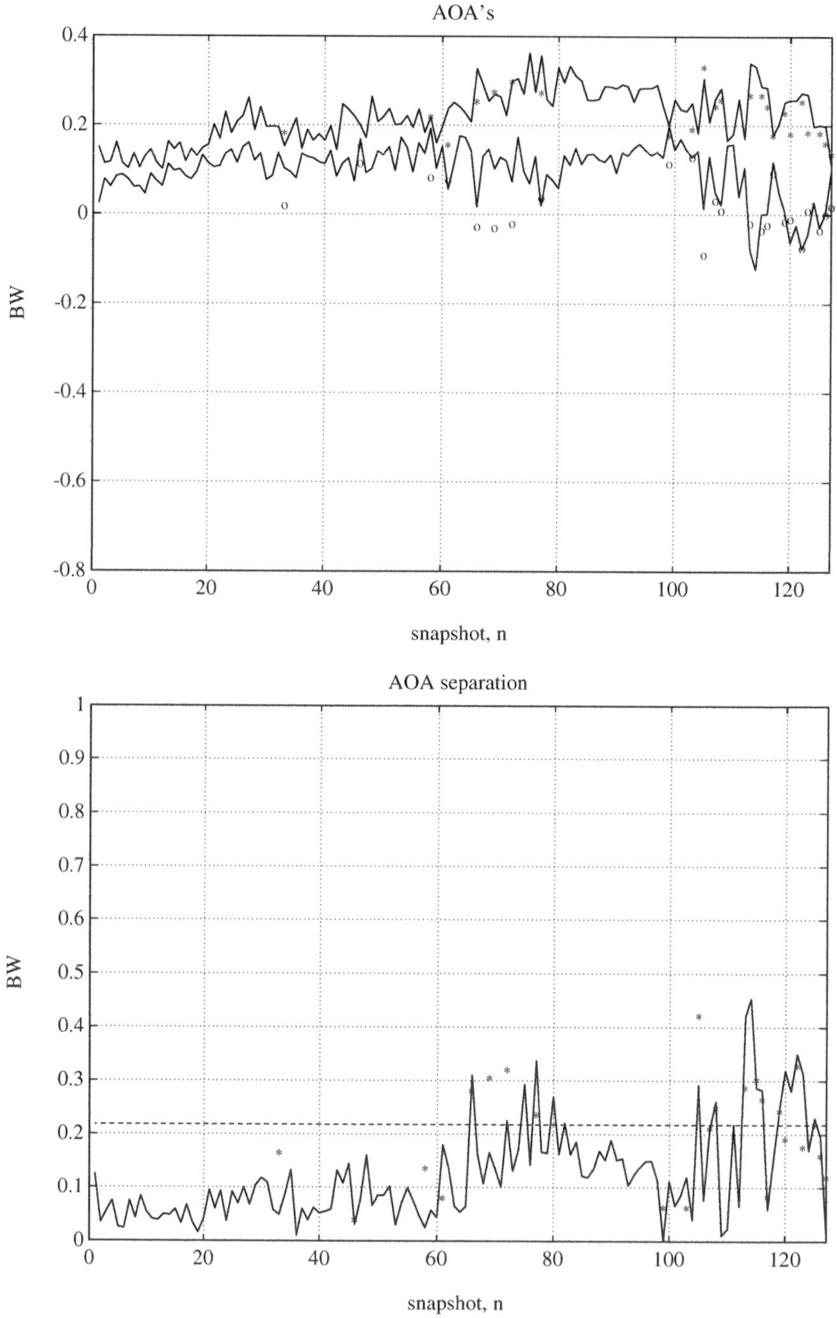

Figure 2.31 Dataset 2, *now3:dh4.dat;3*, $f = 8.62\,\text{GHz}$, far-field calibrated by *nov1:cff6.dat;3*. Solid lines always correspond to MTM estimates. In the top graph the direct component is above the specular, and * and ° correspond to the ML direct and specular components, respectively. The second graph shows the estimated angular separation between direct and specular components compared with the expected one (dashed line).

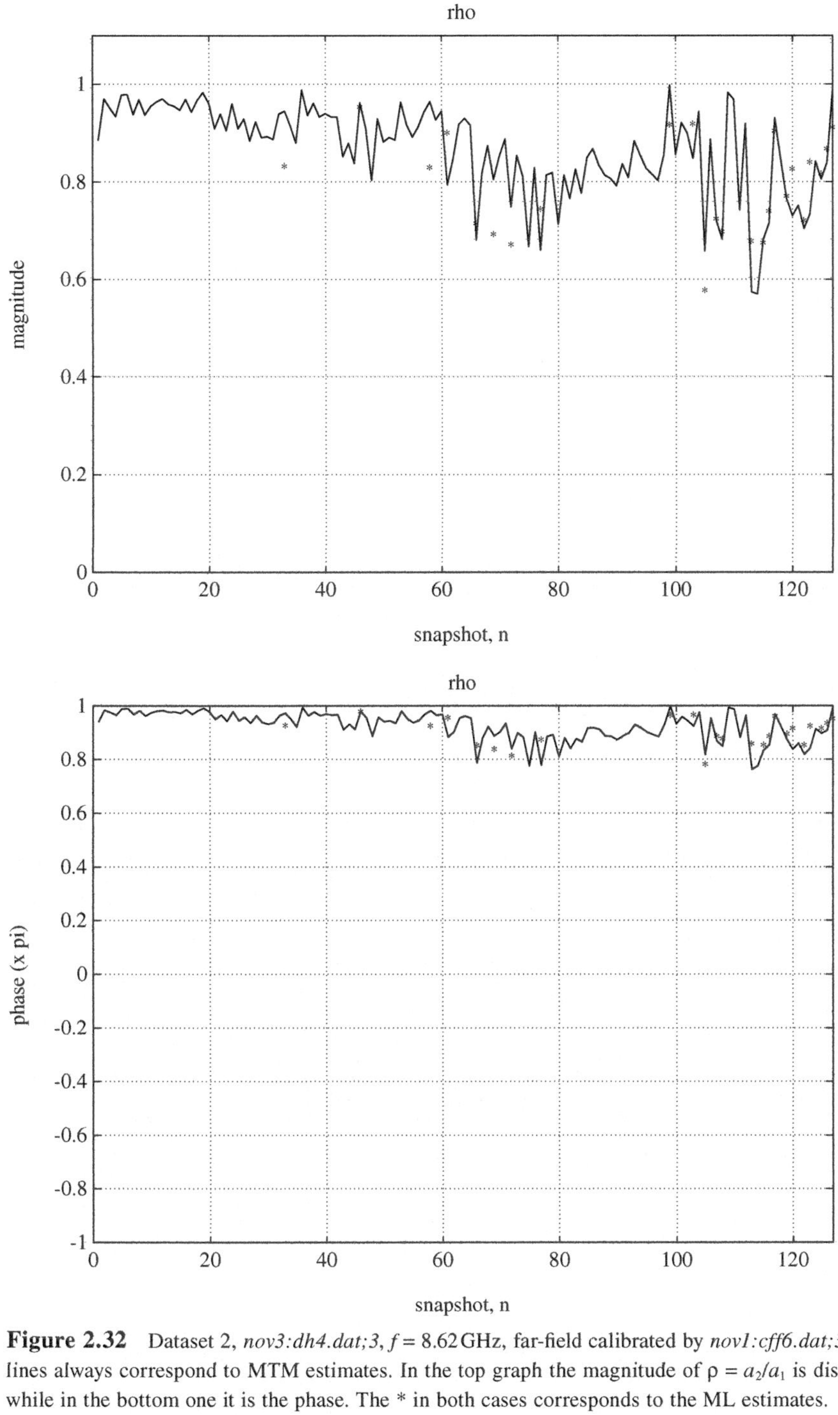

Figure 2.32 Dataset 2, *nov3:dh4.dat;3*, $f = 8.62\,\text{GHz}$, far-field calibrated by *nov1:cff6.dat;3*. Solid lines always correspond to MTM estimates. In the top graph the magnitude of $\rho = a_2/a_1$ is displayed, while in the bottom one it is the phase. The * in both cases corresponds to the ML estimates.

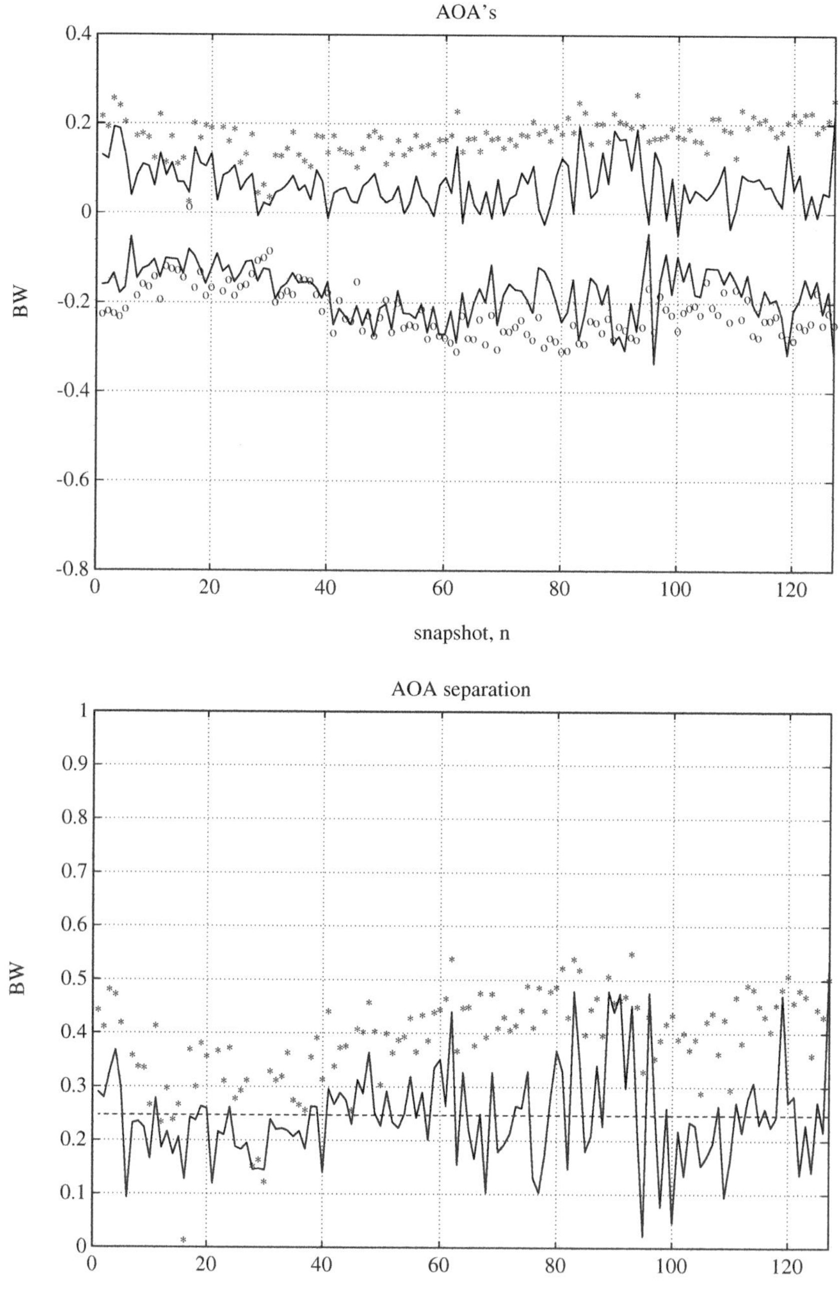

Figure 2.33 Dataset 3, *nov3:dh6.dat;7*, $f = 9.76\,\text{GHz}$, far-field calibrated by *nov1:cff7.dat;7*. Solid lines always correspond to MTM estimates. In the top graph the direct component is above the specular, and * and ° correspond to the ML direct and specular components, respectively. The second graph shows the estimated angular separation between direct and specular components compared with the expected one (dashed line).

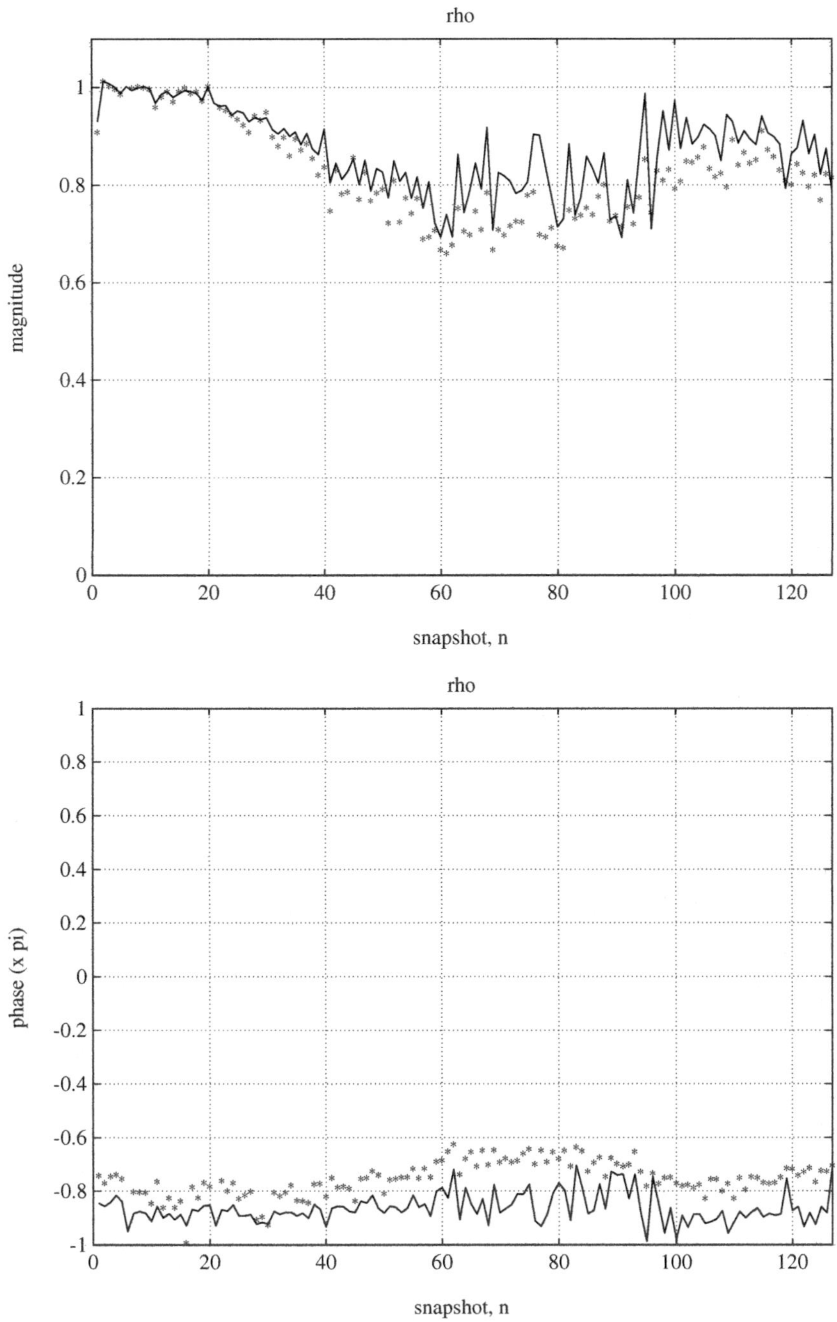

Figure 2.34 Dataset 3, *nov3;dh7,dat;7*, f = 9.76 GHz, far-field calibrated by *novl:cff7.dat;7.* Solid lines always correspond to MTM estimates. In the top graph the magnitude of $\rho = a_2/a_1$ is displayed, while in the bottom one it is the phase. The * in both cases corresponds to the ML estimates.

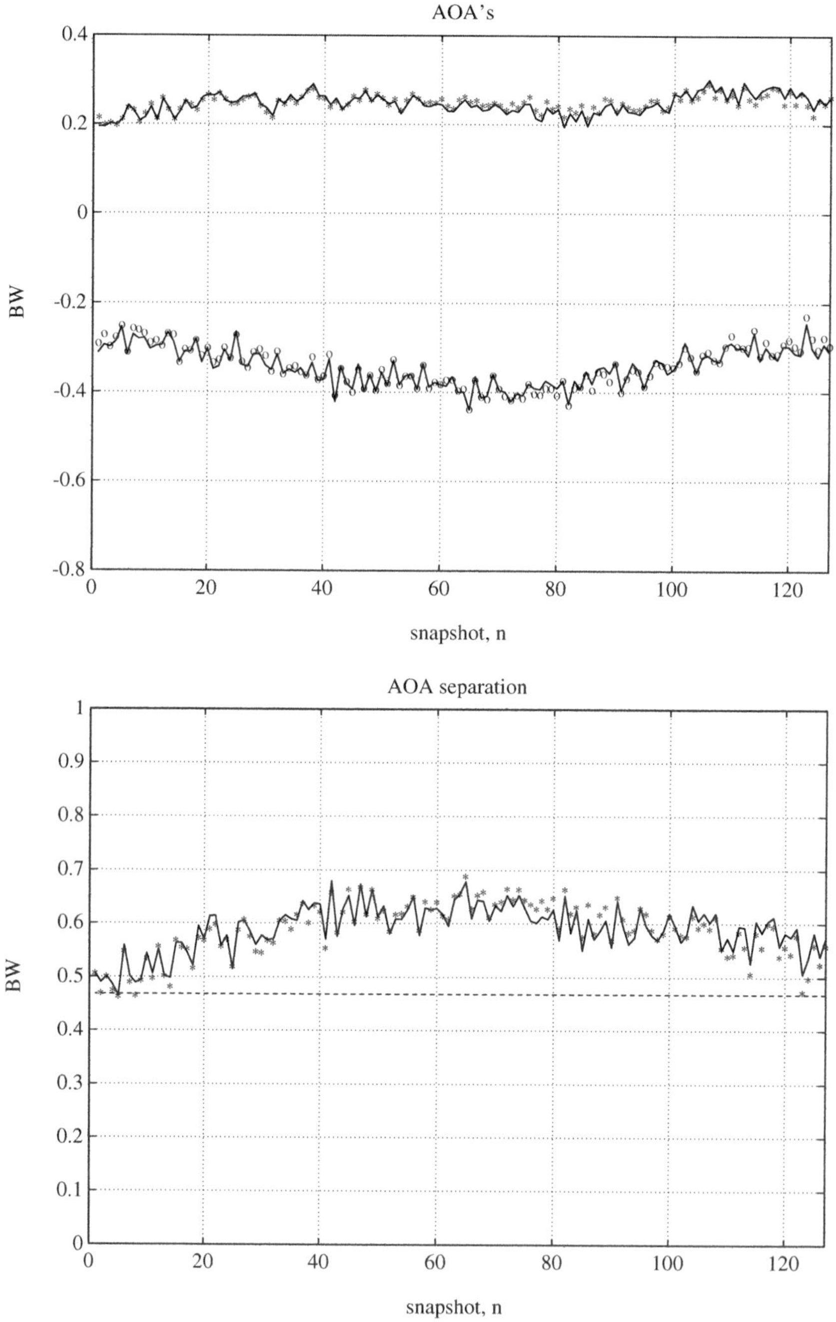

Figure 2.35 Dataset 4, *nov3;dd1.dat;2*, f = 10.12 GHz, far-field calibrated by *nov1:cff6,dat;8*. Solid lines always correspond to MTM estimates. In the top graph the direct component is above the specular, and * and ° correspond to the ML direct and specular components, respectively. The second graph shows the estimated angular separation between direct and specular components compared with the expected one (dashed line).

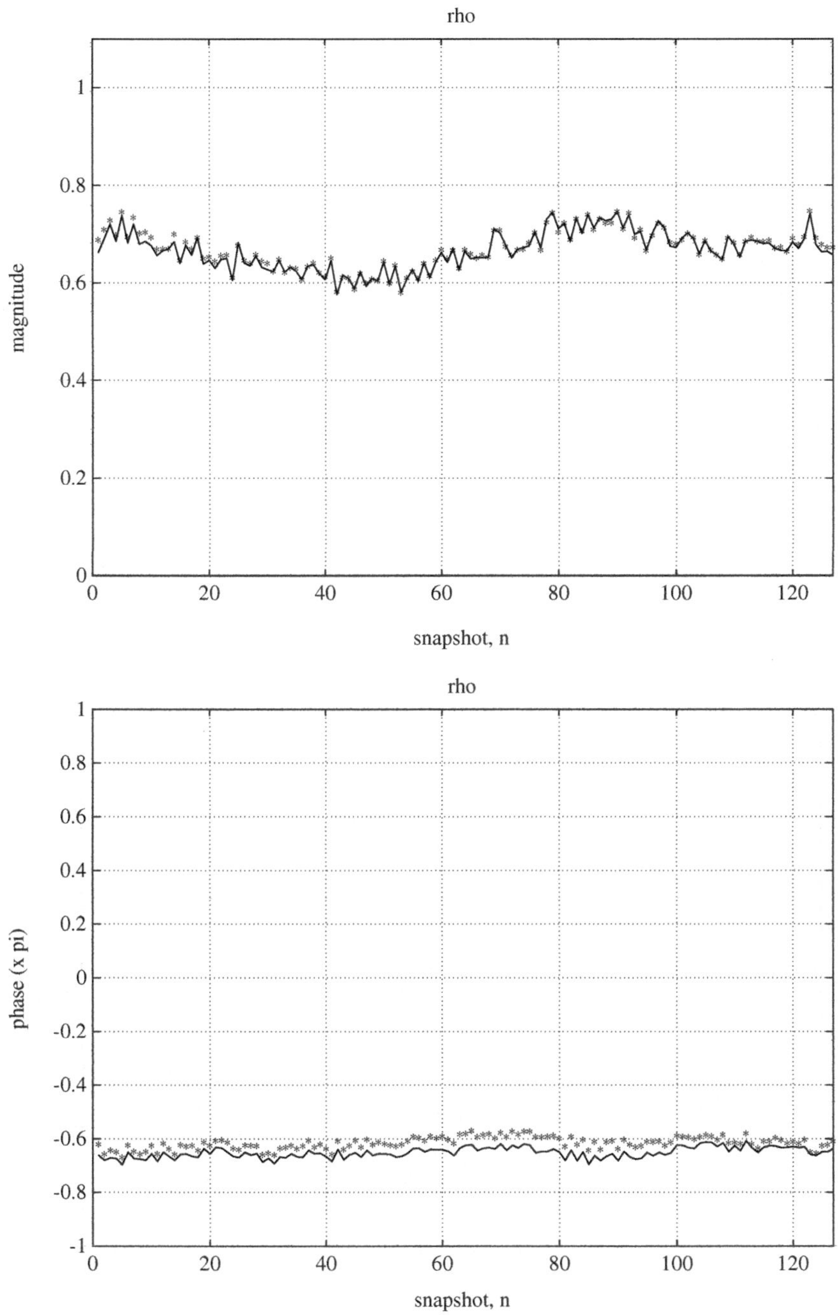

Figure 2.36 Dataset 4, *nov3;dd1.dat;2*, f = 10.12 GHz, far-field calibrated by *nov1:cff6.dat;8*. Solid lines always correspond to MTM estimates. In the top graph the magnitude of $\rho = a_2/a_1$ is displayed, while in the bottom one it is the phase. The * in both cases corresponds to the ML estimates.

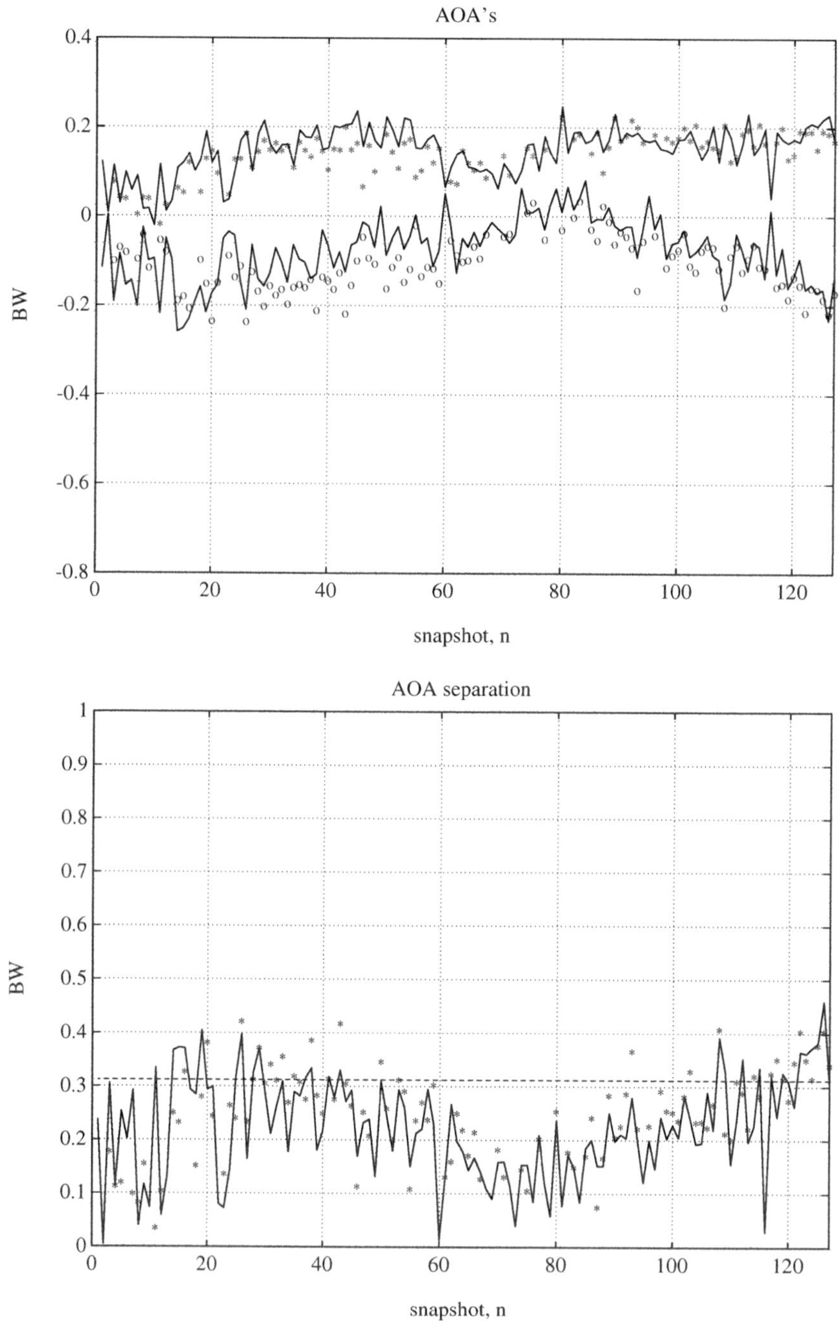

Figure 2.37 Dataset 5, *nov3:dh4.dat;16*, f = 12.34 GHz, far-field calibrated by *nov1:cff6.dat;16*. Solid lines always correspond to MTM estimates. In the top graph the direct component is above the specular, and * and ° correspond to the ML direct and specular components, respectively. The second graph shows the estimated angular separation between direct and specular components compared with the expected one (dashed line).

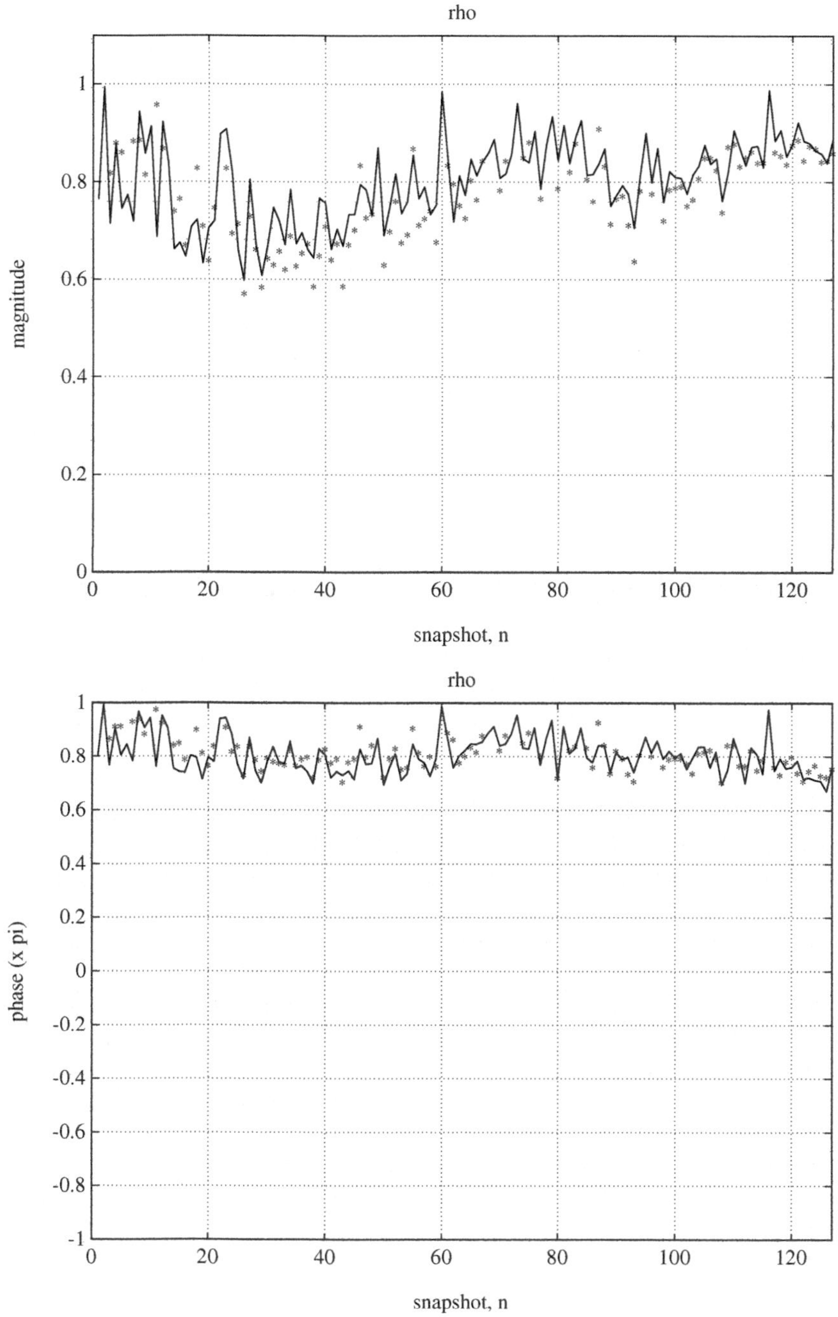

Figure 2.38 Dataset 5, *nov3:dh4.dat;16*, f = 12.34 GHz, far-field calibrated by *nov1:cff6.dat;16*. Solid lines always correspond to MTM estimates. In the top graph the magnitude of $\rho = a_2/a_1$ is displayed, while in the bottom one it is the phase. The * in both cases corresponds to the ML estimates.

also that the estimated ρ is quite small. This means the presence of a weaker in power specular component, which explains the estimation difficulty.

Dataset 2. The situation here appears to the tougher for ML. Convergence occurred at much fewer points than with MTM. We again see in the first half of this dataset, a bias in the AOA separation.

Dataset 3. The behavior of both methods is quite good here, with MTM having a slight edge in the observed AOA separation.

Dataset 4. The bias observed in this case seems to have a periodic behavior. In this situation, the antenna tower was being tilted by the force of the wind during all observation periods, and the periodic variation observed may well have been caused by the tower swaying away from the vertical, making the actual separation angles larger than expected.

Dataset 5. The estimates here seem to be quite noisy but they agree quite well in all cases, and the periodic disturbance in the AOA separation is smaller.

2.10 DIFFUSE MULTIPATH SPECTRUM ESTIMATION

If we remove the estimated plane waves (direct + specular components) from the measured spectrum, the residual spectrum should correspond to the diffuse component. In this way, we can estimate the diffuse component in both the frequency (array element samples in time) and wavenumber (array element samples in space; snapshots) domains. An example of a fully reconstructed wavenumber spectrum is given in Fig. 2.39. This is the first snapshot of *nov3:dh3.dat.2*. Note that as far as we are aware, this is the first time that a full picture of an experimentally measured wavenumber spectrum has been reported for a low-angle tracking radar environment. Both multipath components are displayed, together with the direct signal, and it is clearly seen that assuming a white noise background is wrong.

A variety of different HH subsets of data from the fixed 10.2-GHz channel of *nov3:dh3.dat*, *nov3:dh4.dat* and *nov4:dh9.dat* were also investigated. Their residual frequency and wavenumber spectra are given in Figs. 2.40–2.43. Finally, similar residual spectra are computed from the datasets examined in the previous section (Figs. 2.44–2.49).

In both cases, the spectra display a straightforward behavior. The broad peak observed at baseband can be modeled as a Gaussian spectrum. However, for the wavenumber spectra, the situation appears to be more complex. The first set of spectra (Figs. 2.41 and 2.43) show a pronounced peak in the negative part of the spectrum (corresponding to directions of arrival from below the horizon), while such is not the case for the second set. Several strong peaks are observed in both sides of the spectrum. Assuming for the moment that the behavior depicted is true (see next section), we can now explain some of the observations in the previous section. Specifically, we may make the following observations:

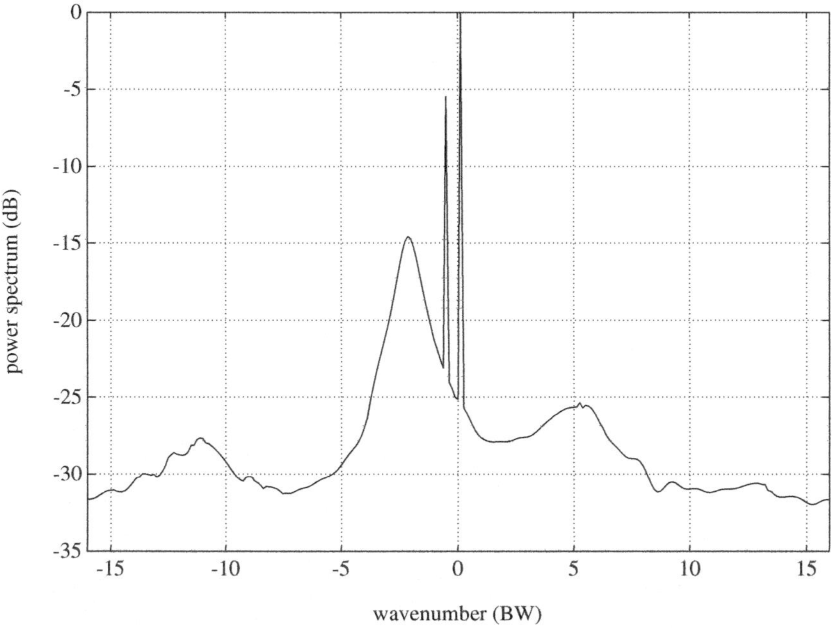

Figure 2.39 The fully reconstructed wavenumber spectrum of the first snapshot of *nov3:dh3.dat;2* far-field calibrated with *nov2:cff2.dat;1*.

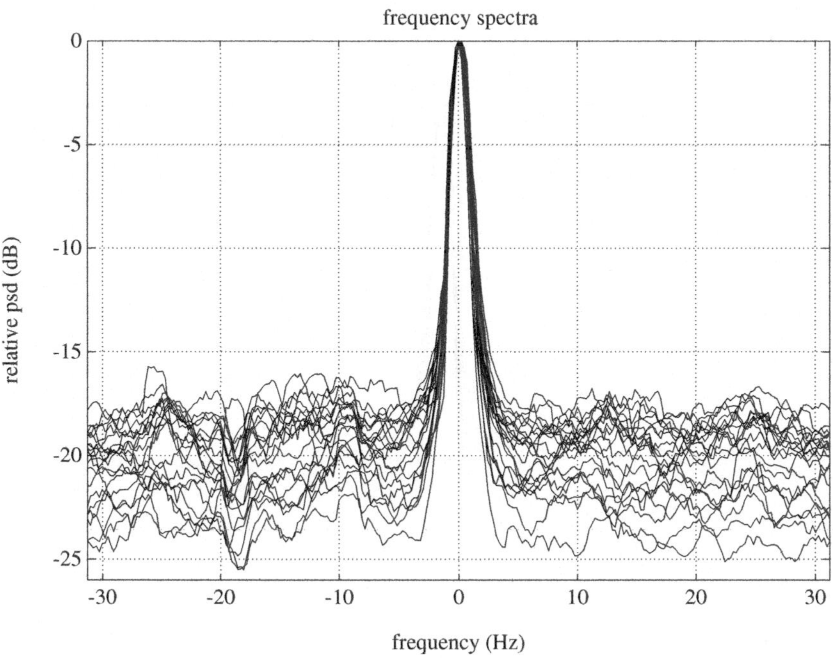

Figure 2.40 The frequency spectra of a variety of similar datasets from *nov3:dh3.dat;** and *nov3: dh4.dat;** all at 10.2 GHz.

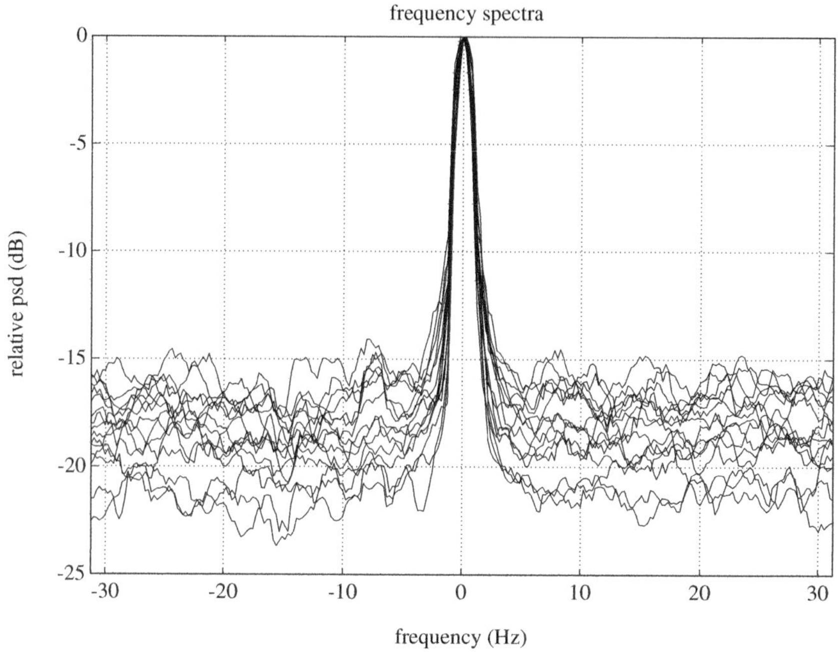

Figure 2.41 The wavenumber spectra of a variety of similar datasets from *nov3:dh3.dat;** and *nov3: dh4.dat;**.

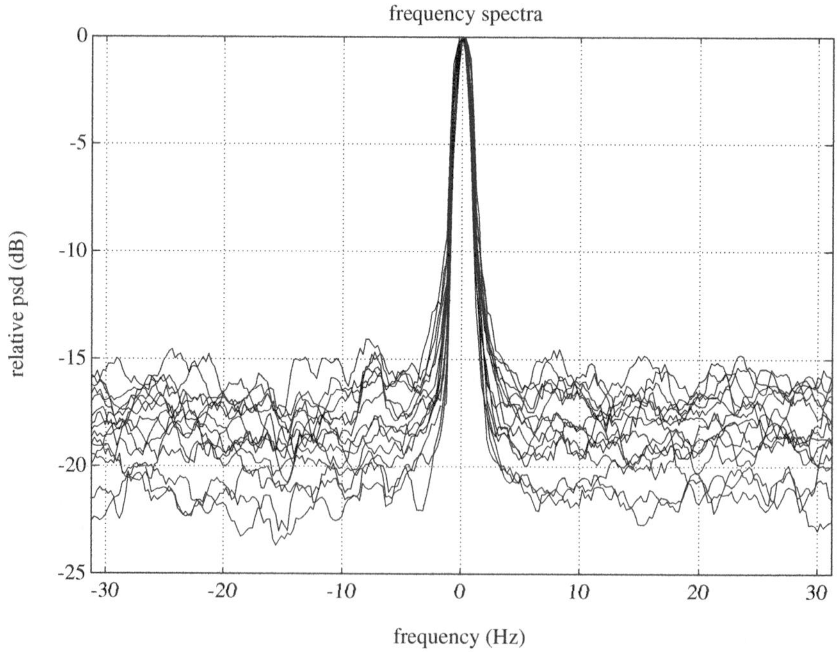

Figure 2.42 The frequency spectra of a variety of datasets from *nov4:dh9.dat;** at 10.2 GHz.

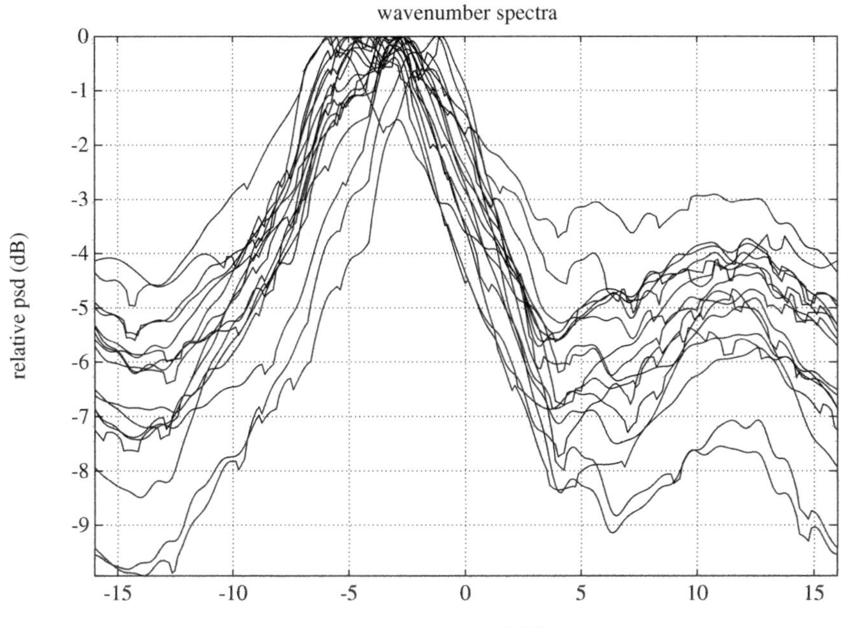

Figure 2.43 The wavenumber spectra corresponding to the frequency spectra of Fig. 2.42. Note that the diffuse peak center is now less well defined, due to the estimated AOA separation being smaller than that for the previous case.

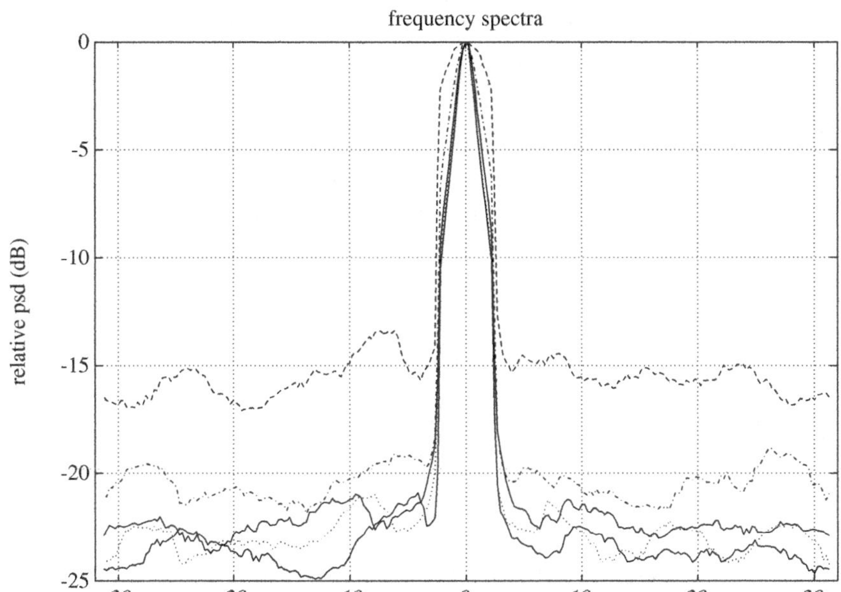

Figure 2.44 Frequency spectra for all five datasets examined in the previous section.

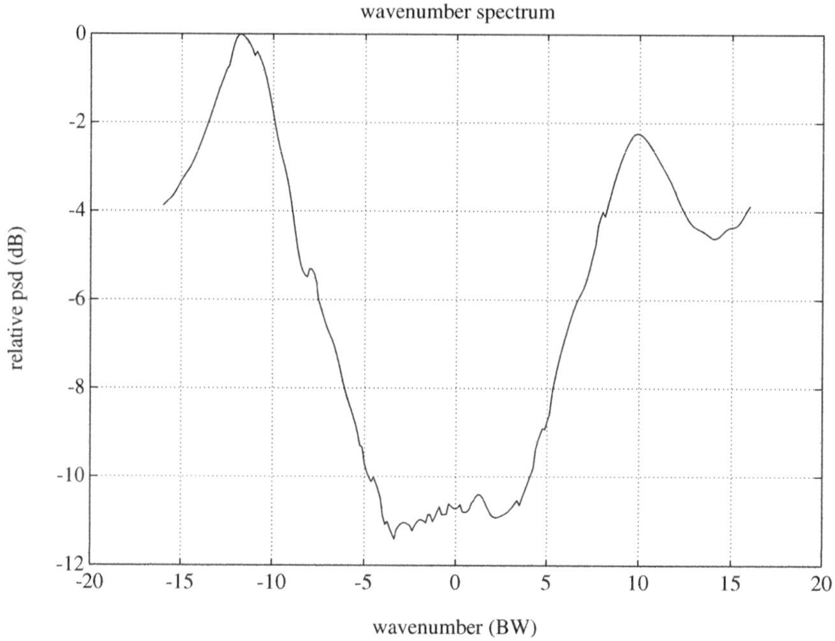

Figure 2.45 Wavenumber spectrum of dataset 1, *nov4:dh9.dat;1.*

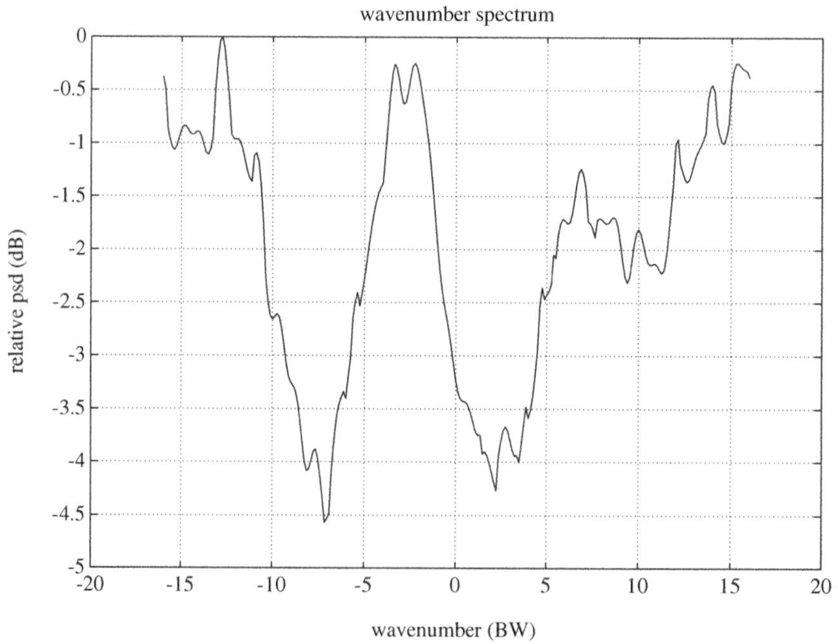

Figure 2.46 Wavenumber spectrum of dataset 2, *nov3:dh4.dat;3.*

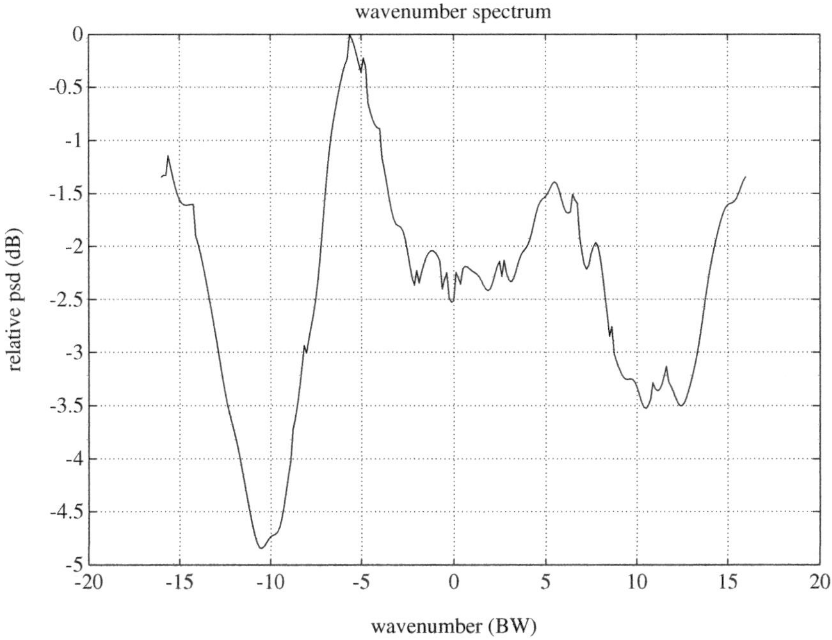

Figure 2.47 Wavenumber spectrum of dataset 3, *nov3:dh6.dat;7.*

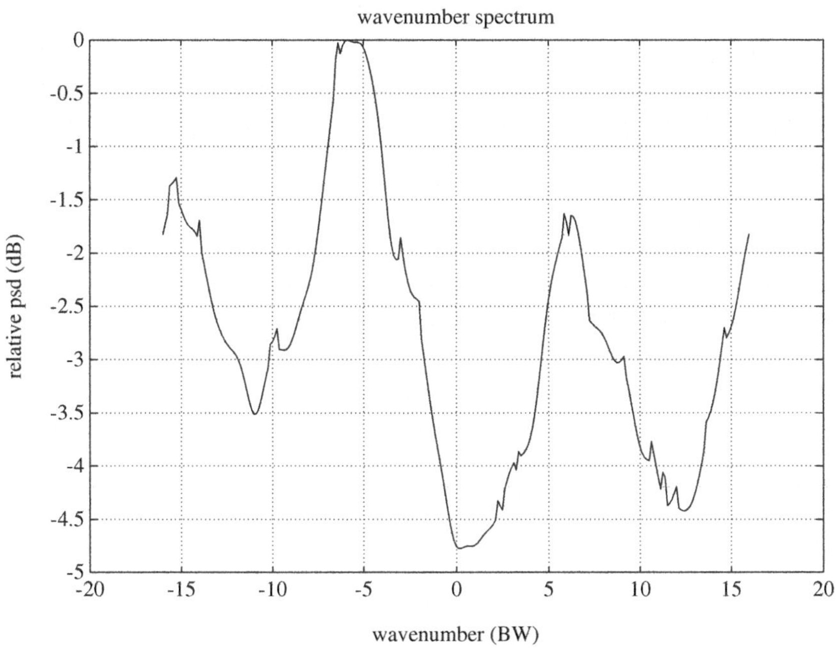

Figure 2.48 Wavenumber spectrum of data set 4, *nov3:dd2.dat;2.*

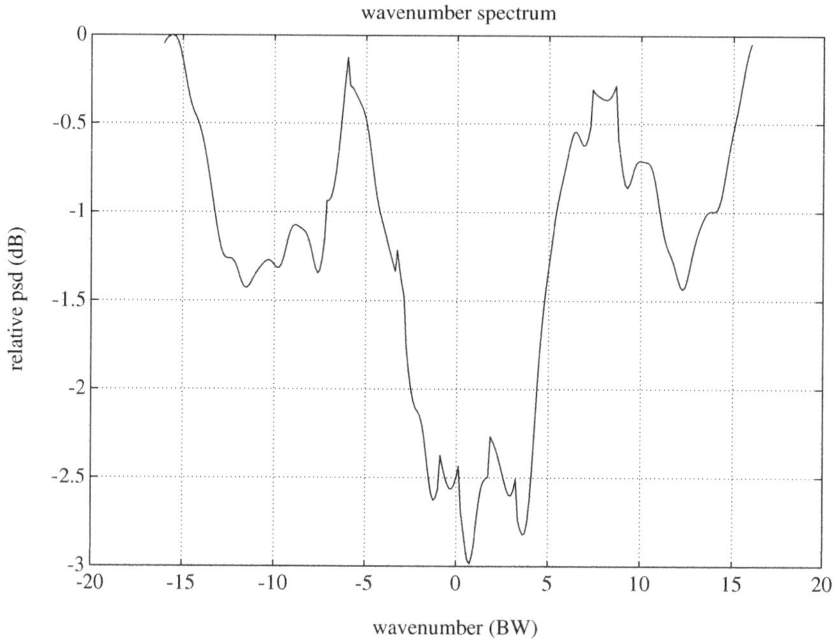

Figure 2.49 Wavenumber spectrum of dataset 5, *nov3:dh4.dat;16*.

Dataset 1. We have two, strong, diffuse peaks appearing in the wavenumber spectrum, which could influence the behavior of the estimated AOAs compared to that actually observed. Note also that the expected separation assumes a specular model. This should be a better approximation as the assumed model mismatch (i.e., weak diffuse component near the direct and specular AOAs) becomes smaller.

Dataset 2. More than one region in the wavenumber domain shows a strong diffuse component. In addition, the frequency-domain spectrum for this dataset displays a higher noise floor by about 5 dB from all other cases; it is therefore no wonder that ML often breaks down here.

Dataset 3. There appears to be a plateau around the origin which, locally at least, is closer to the specular model assumption, so that both the expected and the observed separations agree well enough. Note that MTM, which does not assume a white noise background, performs slightly better.

Dataset 4. Similar behavior to dataset 1 is observed for this dataset, except that the specular component power is larger here.

Dataset 5. Similar behavior to dataset 2 is observed here. However, since the observed AOA separation is slightly larger, the overall behavior is better.

2.11 DISCUSSION

Examining both the estimated AOAs and the diffuse component spectra reported in Sections 2.9 and 2.10 for real-life data, we may now make some final observations:

- The AOA estimate often shows a strong variability during the observation interval, which justifies the snapshot-by-snapshot setup implemented.
- The multi-taper method (MTM), which does not assume a white noise background, always gives AOA estimates with separation larger than 0.001 BW, while the maximum likelihood ML mehtod occasionally does not. The breakdown of ML is more likely to occur when the separation between the direct and specular components is very small (~0.1 BW). Model mismatch in neglecting the diffuse component is then significant. Note, however, that ML can be improved by incorporating additional information in the form of *a priori* knowledge or frequency agility [18, 21]. Similar modifications can also be applied to MTM [9].
- In the context of AOA estimation, there does not seem to be an overwhelming evidence in favor of MTM being superior to ML (provided that ML does not break down). It is encouraging, however, to see that MTM does perform so well compared with ML. However, MTM is the only nonparametric way to get a good estimate of the background diffuse component. Note also that multipath is correlated with itself both spatially and temporally. MTM can be extended in a natural way to obtain coherence estimates between the individual components.

Concerning the shape of the spectra in the frequency domain, they can be modeled as a Gaussian spectrum, the reason being that the contribution of a large number of moving scatterers (law of large numbers) leads to the peak broadening observed.

In the wavenumber domain, positive wavenumber values correspond to the volume space above the physical horizon, while negative values correspond to the space below the horizon, scattering surface included. Two rough surface specular point models were investigated, Barton's [1] and McGarty's [24]. However, only the latter model (when shadowing is included) yields qualitative results similar to some of the experimental cases observed.

Before a comparison can be made between theoretical models and experimentally observed spectra, the theoretically obtained spectra must be modified to take into account the wrapping and grating lobe effects that occur since $d/\lambda > 0.5$ in our experimental setup. The choice of $d/\lambda > 0.5$ was made due to the practical limitation of how closely together the horn antenna elements could be placed. For a frequency of 10.2 GHz, we have $d/\lambda \sim 1.94$. This leads to the correspondence between physical angle θ and wavenumber ϕ seen in Fig. 2.50. The five domains observed should be convolved with the antenna field pattern (array factor $\times$ array element field pattern) in order to obtain the power spectrum that should be observed at the receiver. Figure 2.51 shows a sample case from McGarty's model [24] with a roughness scale s = 0.1 (the ratio of the surface height variance to the surface correlation length) and a geometry similar to that described in Section 2.8.

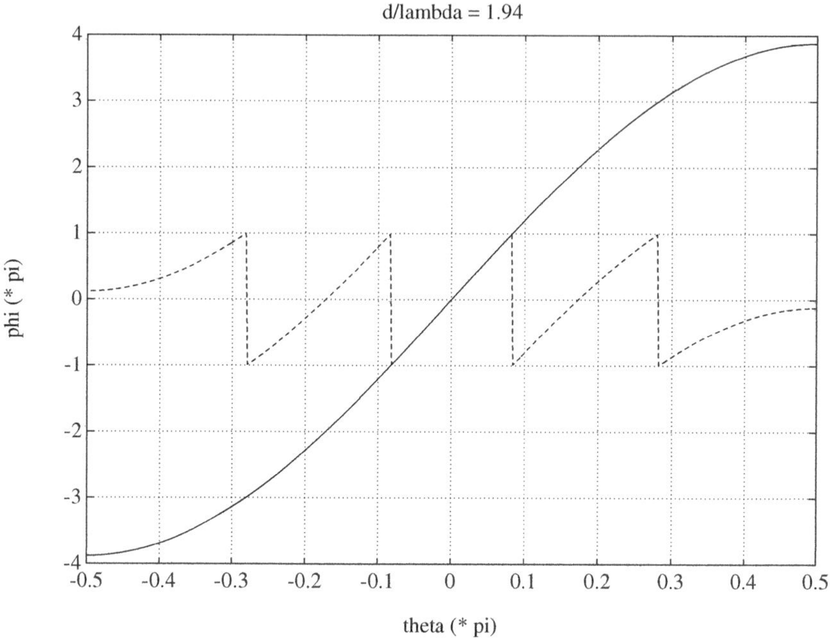

Figure 2.50 The wavenumber ϕ as a function of θ for the case $d/\lambda = 1.94$. The analytic form of the mapping is $\phi = (2\pi d/\lambda)\sin\theta$.

It is only when shadowing is included that we see a broad, diffuse peak between −10 and 0 BW. This agrees with some of the experimentally observed spectra. Most spectra, however, display additional peaks indicating (unless there is a strong calibration problem) other scattering centers, beyond the unambiguous field of view of the antenna receiver array. Returns from these centers would be wrapped into the positive wavenumber part of the spectrum, leading to the observed additional peaks. Since the model employs only a simplified version of shadowing and only a single-scale rough surface, we should not expect perfect agreement. Ideas on sea surface modeling indicate that a coherent, composite model of two roughness scales may be necessary for good predictions.

In conclusion, we would like to highlight the following noteworthy points:

- MTM offers an exceptional approach to the problem of high-resolution full-spectrum estimation. The nonparametric nature of the method in estimating the continuous background spectra (colored noise) is particularly useful when no *a priori* knowledge, or accurate model of the underlying process(es) is available. In fact, this is the only method that performs so well in this kind of situation.
- Regarding AOA estimation, both MTM and ML (maximum likelihood with white noise background) give consistent results for the low-grazing angle case examined in this chapter. However, MTM is found to be slightly superior to ML, although the difference between the two estimators is not overwhelming.

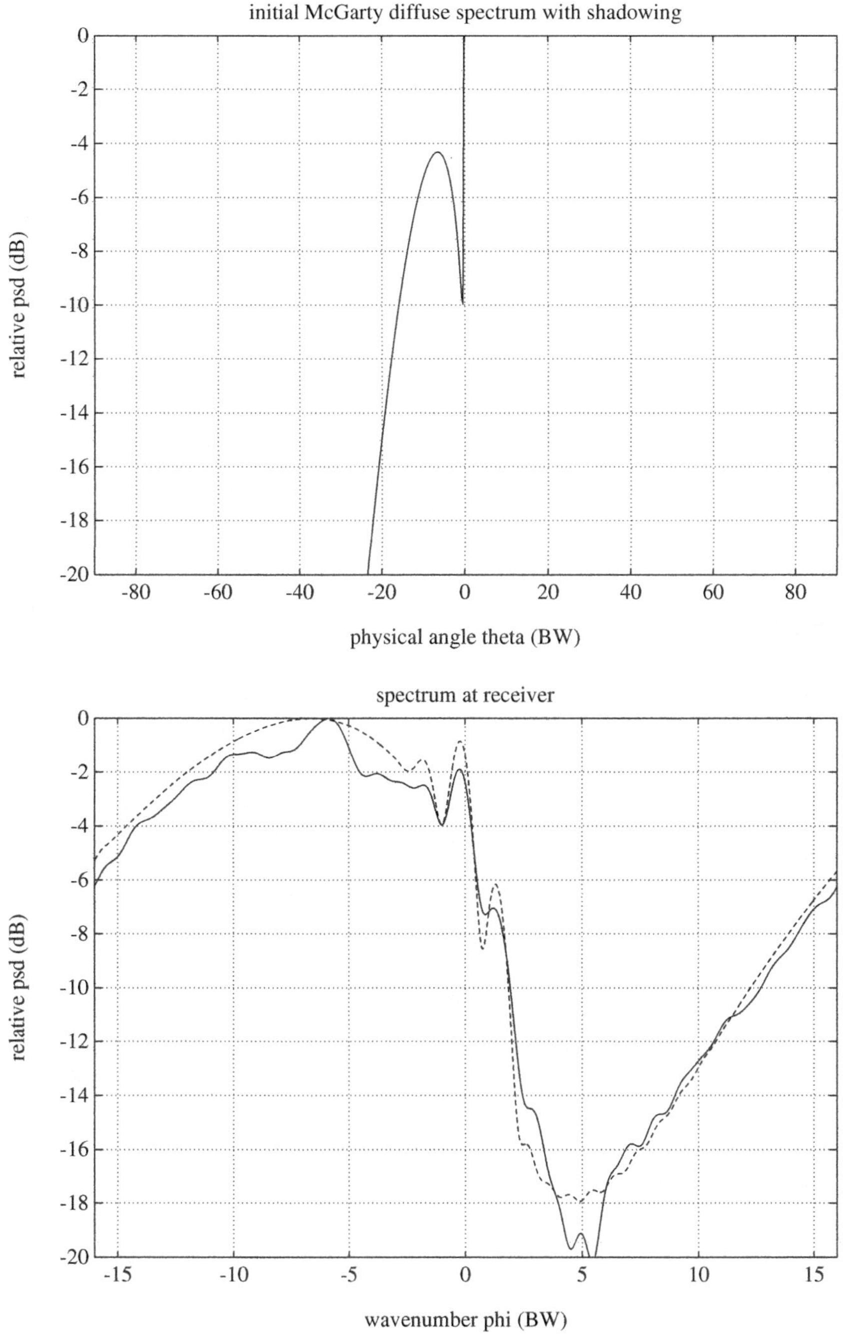

Figure 2.51 The McGarty diffuse spectrum for a geometry similar to ours (top graph). This is transformed in the bottom graph, with wraparound and grating lobes included, and gives the theoretical spectrum (dashed line) that should be observed at the receiver. The solid line graph illustrates the average spectrum of 127 snapshots, generated by a Karhunen–Loève expansion. This is a more likely spectrum shape to be observed at the receiver according to this model.

- The F-tests used for AOA or tone estimation of harmonic components come in two varieties, point and integral regression. Our experience so far suggests that the latter is more stable than the former, although the final estimates in the cases examined here do not differ appreciably. Point regression is also seen to be much faster in execution time on the computer, offering an excellent first glimpse on the harmonic content of a data sequence.

A final point to make in this discussion: The promising experimental results presented herein, which were obtained using an instrument-quality sampled-aperture bistatic radar system and operating under challenging low-grazing angles, lead us to suggest that the multi-taper method (MTM) is an exceptional tool for full-spectrum estimation. It is our belief, based on our experience with it, that is may not be possible to obtain better results with any other nonparametric approach. In fact, MTM is also a better general-purpose spectrum estimator than many parametric ones.

Recognizing the principled theoretical basis of MTM and its general-purpose applicability as a nonparametric spectrum estimator, building a computationally efficient MTM processor capable of operating in real-time for radar applications is indeed a research challenge.

REFERENCES

1. D. K. BARTON (1974). Low-angle radar tracking, *Proc. IEEE* **62**, 687–704.
2. P. BECKMANN AND A. SPIZZICHINO (1987). *The Scattering of Electromagnetic Waves from Rough Surfaces*, Artech House, Norwood, MA (a reprint of the classic 1963 monograph).
3. R. B. BLACKMAN AND J. W. TUKEY (1959). *The Measurement of Power Spectra*, Dover, New York, (a reprint of papers appearing in the January and March issues of the B.S.T.J. in 1958).
4. L. V. BLAKE (1986). *Radar Range-Performance Analysis*, Artech House, Norwood, MA.
5. T. P. BRONEZ (1988). "Nonparametric spectral estimation of irregularly sampled multidimensional random processes," Ph.D. dissertation, Arizona State University, Tempe, AZ.
6. T. P. BRONEZ (1988). Spectral estimation of irregularly sampled multidimensional processes by generalized prolate spheroidal sequences, *IEEE Trans. Acoust., Speech, Signal Process.* **36**, 1862–1873.
7. N. R. DRAPER AND H. SMITH (1981). *Applied Regression Analysis*, Wiley, New York.
8. A. DROSOPOULOS (1991). Investigation of diffuse multipath at low-grazing angles, Ph.D. dissertation, McMaster University, Department of Electrical Engineering, Hamilton, Ontario, Canada.
9. A. DROSOPOULOS AND S. HAYKIN (1991). Angle-of-Arrival Estimation in the Presence of multipath, *Electronic Lett.* **27**(10), 798–799.
10. S. HAYKIN (1985). Radar array processing for angle of arrival estimation, in *Array Signal Processing*, S. HAYKIN (ed.), Prentice-Hall, Englewood Cliffs, NJ, Chapter 4.
11. R. HOOKE AND T. A. JEEVES (1961). Direct search solution of numerical and statistical problems, *J. ACM* **8**, 212–229.
12. A. ISHIMARU (1978). *Wave Propagation and Scattering in Random Media*, Vol. II, Academic Press, New York.
13. J. O. JONSSON AND A. O. STEINHARDT (1993). The total probability of false alarm of the multi-window harmonic detector and its application to real data, *IEEE Trans. Signal Processing* **41**(4), 1702–1705.
14. A. F. KAUPE. Algorithm 178: Direct search, Collected algorithms from *Commun. ACM* **178**-P1R1.
15. S. M. KAY AND S. L. MARPLE, JR. (1981). Spectrum analysis—a modern perspective, *Proc. IEEE* **69**, 1380–1419.

16. D. E. KERR (ed.) (1951). *Propagation of Short Radio Waves. M.I.T. Radiation Laboratory Series*, Vol. 13, McGraw-Hill, New York.
17. V. KEZYS, personal communication.
18. V. KEZYS AND S. HAYKIN (1988). Multi-frequency angle-of-arrival estimation: An experimental evaluation, Proc. SPIE, *Advanced Algorithms and Architectures for Signal Processing III* **975**, 93–100.
19. B. KLEINER, R. D. MARTIN, AND D. J. THOMSON (1979). Robust estimation of power spectra, *J.R. Statist. Soc. B* **41**, 313–351.
20. C. R. LINDBERG AND J. PARK (1987). Multiple-taper spectral analysis of terrestrial free oscillations: Part II, *Geophys. J.R. Astron. Soc.* **91**, 795–836.
21. T. LO AND J. LITVA (1991). Use of a highly deterministic multipath signal model in low-angle tracking, *IEE Proc.*, Part F, **138**, 163–171.
22. S. L. MARPLE, JR. (1987). *Digital Spectral Analysis with Applications*, Prentice-Hall, Englewood Cliffs, NJ.
23. R. D. MARTIN AND D. J. THOMSON (1982). Robust-resistant spectrum estimation, *Proc. IEEE* **70**, 1097–1115.
24. T. P. MCGARTY (1976). Antenna performance in the presence of diffuse multipath, *IEEE Trans. Aerosp. Electron. Systems* **12**, 42–54.
25. C. T. MULLIS AND L. L. SCHARF (1991). Quadratic estimators of the power spectrum, in *Advances in Spectrum Analysis and Array Processing*, Vol. I, S. Haykin (ed.), Prentice-Hall, Englewood Cliffs, NJ, Chapter 1.
26. R. ONN AND A. O. STEINHARDT (1991). A multi-window method for spectrum estimation and sinusoid detection in an array environment, *Proc. SPIE, Advanced Algorithms and Architectures for Signal Processing*, San Diego.
27. A. PAPOULIS (1984). *Probability, Random Variables and Stochastic Processes*, 2nd edition, McGraw-Hill, New York.
28. J. PARK (1987). Multitaper spectral analysis of high-frequency seismographs, *J. Geophys. Res.* **92**(B12), 12675–12684.
29. J. PARK, C. R. LINDBERG, AND D. J. THOMSON (1987). Multiple-taper spectral analysis of terrestrial free oscillations: Part I, *Geophys. J.R. Astron, Soc.* **91**, 755–794.
30. W. H. PRESS, B. P. FLANNERY, S. A. TEUKOLSKY, AND W. T. VETTERLING (1986). *Numerical Recipes*, Cambridge University Press, New York.
31. D. SLEPIAN (1965). Some asymptotic expansions for prolate spheroidal wave functions, *J. Math, Phys.* **44**, 99–140.
32. D. SLEPIAN (1968). A numerical method for determining the eigenvalues and eigenfunctions of analytic kernels, *SIAM J. Numer. Anal.* **5**, 586–600.
33. D. SLEPIAN (1978). Prolate spheroidal wave functions, Fourier analysis, and uncertainty—V: The discrete case, *Bell System Tech. J.* **57**, 1371–1430.
34. D. SLEPIAN AND E. SONNENBLICK (1965). Eigenvalues associated with prolate spheroidal functions of zero order, *Bell System Tech. J.* **44**, 1745–1760.
35. D. J. THOMSON (1977). Spectrum estimation techniques for characterization and development of WT4 waveguide, *Bell System Tech. J.* **56**, 1769–1815, 1983–2005.
36. D. J. THOMSON (1982). Spectrum estimation and harmonic analysis, *Proc. IEEE* **70**, 1055–1096.
37. D. J. THOMSON (1989). Multiple-window bispectrum estimation, in *Proc. Workshop on Higher-Order Spectral Analysis*, Vail, CO, pp. 19–23.
38. D. J. THOMSON (1990). Quadratic-inverse spectrum estimation: Applications to paleoclimatology, *Phil. Trans. R. Soc. London, Ser. A* **332**, 539–597.
39. D. J. THOMSON (1990). Time series analysis of Holocene climate data, *Phil. Trans. R. Soc. London, Ser. A* **330**, 601–616.
40. D. J. THOMSON AND A. D. CHAVE (1991), Jackknifed error estimates for spectra, coherences and transfer functions, in *Advances in Spectrum Analysis and Array Processing*, Vol. I, S. HAYKIN (ed.), Prentice-Hall, Englewood Cliffs, NJ, Chapter 2.
41. A. M. YAGLOM (1987). *Correlation Theory of Stationary and Related Random Functions I: Basic Results*, Springer-Verlag, New York.

Chapter 3

Time–Frequency Analysis of Sea Clutter†

David J. Thomson and Simon Haykin

3.1 INTRODUCTION

In Chapter 2, we studied a novel application of the multi-taper method as a tool for wavenumber spectrum estimation in a low-tracking radar environment, which is characterized by the combined presence of specular and diffuse multipath. The multipath studied therein focused on a sea surface with the target signal originating at low grazing angles. In this chapter, we use an expanded form of the multi-taper method to study *sea clutter*, which refers to the radar backscatter from an ocean (or sea) surface. Sea clutter is of particular concern in the detection of a target (particularly a small target), a task that is made difficult by the presence of sea clutter.

Typically, sea clutter is *nonstationary* in that its statistical characterization varies with time. The power spectrum of sea clutter is therefore a function of two variables: time and frequency. To accommodate this new situation, we need to expand our radar signal-processing horizon by focusing on time–frequency analysis, which is indeed the objective of this chapter.

The chapter is organized as follows. Section 3.2 presents background theory on statistical analysis of nonstationary signals, the roots of which may be traced to seminal work published by Loève in 1946. Section 3.3 describes the Loève spectrum, which includes the well-known Wigner–Ville distribution as a special case. Section 3.4 expands on the material presented on the Loève spectrum by incorporating into it multi-taper spectral estimates. With the theoretical framework described in Sections 3.3 and 3.4 at hand, experimental results on time-frequency analysis of sea clutter, using real-life radar data, are presented in Section 3.5. The chapter concludes in Section 3.6.

† The material presented in this chapter is based partly on the paper: S. Haykin and D. J. Thomson (1998). Signal detection in a nonstationary environment reformulated as an adaptive pattern classification problem, *Proc. IEEE Special Issue on Intelligent Signal Processing* **86**(11), 2325–2344.

3.2 AN OVERVIEW OF NONSTATIONARY BEHAVIOR AND TIME–FREQUENCY ANALYSIS

The statistical analysis of nonstationary signals has had a rather mixed history. Although the general second-order theory was published during 1946 by Loève [1, 2], it has not been applied nearly as extensively as the theory of stationary processes published only slightly previously by Wiener and Kolmogorov. There were, at least, four distinct reasons for this neglect:

1. Loève's theory was *probabilistic*, not statistical, and there does not appear to have been successful attempts to find a statistical version of the theory until some time later.
2. At the time of Loève's publications, the mathematical training of most engineers and physicists in signals and random processes was minimal; and recalling that even Wiener's delightful book was referred to as "The Yellow Peril," it is easy to imagine the reception that a general nonstationary theory would have received.
3. Even if the theory had been commonly understood at the time and good statistical estimation procedures had been available, the computational burden would probably have been overwhelming. This was the era when Blackman–Tukey estimates of the stationary spectrum were developed, not because they were great estimates but, primarily, because they were simple to understand in mathematical terms and computationally more efficient than other forms.
4. Finally, it cannot be denied that the general theory was significantly harder to grasp than that for stationary processes.

Nonetheless, it was realized that many, perhaps most, of the signals being worked with were *nonstationary*; and starting with the available tools (i.e., the ability to estimate the spectrum of a stationary signal), the spectrogram was developed, and, in fact, predated Loève's work; see references 3 and 4. The idea was that if the process is not "too" nonstationary, then for a relatively short-time block a "quasi-stationary" approximation can be used, so that for the length of the block the spectrum can be approximated by its average. It was also recognized that a major drawback of the spectrogram is that the block lengths and offset between blocks are arbitrary. Thus, although speech, underwater sound, radar, and similar communities have much empirical experience to guide such choices, little can be done with a new, possibly unique, data series except "cut and try" methods. Consequently, it is common to regard the spectrogram as a heuristic or ad-hoc method.

To account for the nonstationary behavior of a signal, we have to include *time* (implicitly or explicitly) in a description of the signal. Given the desirability of working in the frequency domain for well-established reasons, we may include the effect of time by adopting a time–frequency description of the signal. In the past many years, many papers have been published on various estimates of time–frequency distributions; see, for example, Cohen's book [5] and the references therein. In most of this work, the signal is assumed to be *deterministic*. In addition, many of the proposed estimators are constrained to match time and frequency *marginal* density conditions. To be specific, let $D(t, f)$ denote a time–frequency distribution of a signal $x(t)$, and it is required that the time marginal satisfy the condition

$$\int_{-\infty}^{\infty} D(t, f)\,df = |x(t)|^2$$

and, similarly, if $y(f)$ is the Fourier transform of $x(t)$, the frequency marginal density must satisfy the second condition

$$\int_{-\infty}^{\infty} D(t, f)\,dt = |y(f)|^2$$

where t denotes continuous time and f denotes frequency. Given the large differences observed between waveforms collected on sensors spaced short distances apart (see, for example, Vernon [6]), the time marginal requirement is a rather strange assumption. Worse, the frequency marginal distribution is, except for a factor of $1/N$, just the periodogram of the signal. It has been known since before the first periodogram was computed, the periodogram is badly biased and inconsistent.[1] Thus we do not consider matching marginal distributions, as commonly defined, to be important.

Similarly, several estimates have been proposed that attempt to reduce the cross-terms[2] in the Wigner–Ville distribution (defined later in the section) by using the analytic signal instead of the original data. However, because the analytic signal is commonly derived by Fourier transforming the data, discarding the negative frequency components, and taking the inverse Fourier transform, the frequency-domain bias of the analytic signal is dominated by the periodogram bias. The opinion has been expressed that these concerns apply only in a near-pathological dataset and that the sidelobe performance of the Slepian sequences is rarely needed. We consider this opinion to be ill-advised, because we rarely know in advance what sidelobe performance is needed. (Slepian sequences were discussed in Chapter 2 and, for convenience of presentation, they are redefined later in this section.) As an example, the dynamic range of the spectrum for the radar data used in this paper exceeds 10^4, so an estimate constrained to match periodogram marginals could easily be in error by an order of magnitude over most of the frequency domain.

This being said, we may make the following two statements:

1. As we describe it below, the expected value of the Wigner–Ville distribution is just a coordinate rotation of the Loève spectrum. It is not, however, a particularly good *statistical* estimator.
2. The basic Wigner–Ville distribution is a *sufficient statistic* in that it can be inverted to recover the original data to within a phase constant [11]. Thus,

1 An inconsistent estimate is one where the variance of the estimate does not decrease with sample size. The first proof of the inconsistency of the periodogram seems to be Rayleigh [7], with Rayleigh [8] and particularly Rayleigh [9] clarifying the ideas. However, Rayleigh did not use the term "inconsistent"; rather, it was not introduced as a statistical term until Fisher's famous paper [10].

2 *Cross-terms* arise when the Wigner–Ville distribution is applied to the sum of two signals. The Wigner–Ville distribution of such a sum is not equal to the sum of the Wigner–Ville distributions of the two signals; the difference is accounted for by the cross-terms.

although it is not an attractive estimate from a statistical viewpoint, completeness properties of the Wigner-Ville distribution allow it to be effective in some applications.

3. The sea clutter and target-plus-clutter data used here are produced by a coherent radar and therefore are complex-valued, so there is no need to estimate the analytic signal.
4. The cross-terms are visually distinctive and thus may be a significant help in recognizing that more than one component is present in the received signal.

In the final analysis, whether we adopt the stochastic or deterministic approach to time–frequency analysis for representing the nonstationary behavior of a signal depends on details of the problem of interest. It is easy to imagine problems where one or the other of these two approaches would be preferable, but there are other problems, such as the radar data used in our examples, where a good case can be made for both viewpoints, depending on the issue of interest.

3.3 THEORETICAL BACKGROUND ON NONSTATIONARITY

Suppose we are given data consisting of a single finite realization of N contiguous samples of a discrete-time process $x(t)$ for $t = 0, \ldots, N - 1$; henceforth, t denotes discrete time. (This notation differs from that used in Chapter 2.) We assume that the process is *harmonizable* (Loève [2]), so that it has the *Cramér,* or *spectral, representation*:

$$x(t) = \int_{-1/2}^{1/2} e^{j2\pi\nu t} dX(\nu) \tag{3.1}$$

where $dX(\nu)$ is the *increment process*. In this chapter, we also assume that the process has zero mean, that is, $\mathbf{E}\{dX(\nu)\} = 0$, and, correspondingly, $\mathbf{E}\{x(t)\} = 0$. (Note that, strictly speaking, this is *not* the same as assuming that an average has been subtracted from the data.) As parameters of interest, we define the *Loève transform* by the covariance function

$$\begin{aligned}\Gamma_L(t_1, t_2) &= \mathbf{E}\{x(t_1)x^*(t_2)\} \\ &= \int_{-\infty}^{\infty}\int_{-\infty}^{\infty} e^{j2\pi(t_1 f_1 - t_2 f_2)} \gamma_L(f_1, f_2) df_1 df_2\end{aligned} \tag{3.2}$$

and the generalized spectral density

$$\gamma_L(f_1, f_2) df_1 df_2 = \mathbf{E}\{dX(f_1) dX^*(f_2)\} \tag{3.3}$$

where * indicates complex conjugate. Equation (3.3) describes the essential feature of nonstationary processes—namely, that there is correlation between different frequencies.

If the process is stationary, by definition, the covariance $\Gamma_L(t_1, t_2)$, depends only on the time difference $t_1 - t_2$, is which case, the Loève spectrum $\gamma_L(f_1, f_2)$ becomes $\delta(f_1 - f_2)S(f_1)$, where $S(f)$ is the ordinary power spectrum. Similarly, for a white nonstationary process, the covariance function becomes $\delta(t_1 - t_2)\, P(t_1)$, where $P(t)$ is the expected power at time t. Thus, as both the spectrum and covariance functions include delta function discontinuities in simple cases, neither should be expected to be "smooth"; and continuity properties depend on direction in the (f_1, f_2) or (t_1, t_2) plane. These problems are more easily dealt with by rotating both the time and frequency coordinates of the generalized correlations (3.2) and spectral densities (3.3), respectively, by 45°. In the time domain, we define the new coordinates to be a "center" t_0 and a delay τ, as shown here:

$$\begin{aligned} t_1 + t_2 &= 2t_0 \\ t_1 - t_2 &= \tau \end{aligned} \tag{3.4}$$

Equivalently, we may write

$$\begin{aligned} t_1 &= t_0 + \tau/2 \\ t_2 &= t_0 - \tau/2 \end{aligned}$$

We denote the covariance function in the rotated coordinates by $\Gamma(\tau, \tau_0)$ and thus write

$$\Gamma_L(t_1, t_2) = \Gamma\left(t_1 - t_2, \frac{t_1 + t_2}{2}\right) = \Gamma(\tau, t_0) \tag{3.5}$$

Similarly, we define new frequency coordinates f and g by writing

$$\begin{aligned} f_1 + f_2 &= 2f \\ f_1 - f_2 &= g \end{aligned} \tag{3.6}$$

Equivalently, we may write

$$f_1 = f + (g/2)$$

Denote the *rotated spectrum* by

$$\gamma(g, f) = \gamma_L\left(f + \frac{g}{2}, f - \frac{g}{2}\right) \tag{3.7}$$

Substituting these definitions in (3.2) shows that the term $t_1 f_1 - t_2 f_2$ in the exponent of the Fourier transform becomes $(t_0 g + \tau f)$, so we write

$$\Gamma(t_0, \tau) = \int_{-\infty}^{\infty}\int_{-\infty}^{\infty} e^{j2\pi(\tau f + t_0 g)}\gamma(g, f)\, df dg \tag{3.8}$$

Because f is associated with the time difference τ, it corresponds to the ordinary frequency of stationary processes and we refer to it as the "stationary" frequency. Similarly, because g is associated with the average time t_0, it describes the behavior of the spectrum over long time spans and we refer to g as the "nonstationary" frequency.

Consider next the continuity of γ as a function of f and g. On the line $g = 0$, the generalized spectral density γ is just the ordinary spectrum with the usual continuity (or lack thereof) conditions normally applying to stationary spectra. As a function of g, however, we expect to find a δ-function discontinuity at $g = 0$ if for no other reason that almost all data contain some stationary additive noise. Consequently, smoothers in the (f, g) plane (or, equivalently, the (f_1, f_2) plane) should not be isotropic, but require much higher resolution along the nonstationary frequency coordinate g than along the ordinary frequency axis f.

A slightly less arbitrary way of handling the g coordinate is to Fourier transform $\gamma(g, f)$ with respect to the nonstationary frequency, g, and define it as the theoretical *dynamic spectrum* of the process. The motivation is to transform the very rapid variation expected around $g = 0$ into a slowly varying function of t_0 while leaving the usual dependence on f. From Fourier transform theory, we know that δ functions in the frequency-domain transform into a constant in the time domain. It follows therefore that, in a stationary process, $D(t_0, f)$ does not depend on t_0 and assumes the simple form $S(f)$. Thus writing

$$D(t_0, f) = \int_{-\infty}^{\infty} e^{j2\pi\tau f} \mathbf{E}\left\{x\left(t_0 + \frac{\tau}{2}\right)x^*\left(t_0 - \frac{\tau}{2}\right)\right\} d\tau \tag{3.9}$$

and recognizing that [5]

$$W(t_0, f) = \int_{-\infty}^{\infty} e^{j2\pi\tau f} x\left(t_0 + \frac{\tau}{2}\right)x^*\left(t_0 - \frac{\tau}{2}\right) d\tau \tag{3.10}$$

as the formula for the Wigner–Ville distribution of the signal $x(t)$, we now see that *the rotated Loève spectrum is the expected value of the Wigner–Ville distribution.* This relation has been rediscovered several times (see, for example, reference 13). Note carefully, however, that unlike the Wigner–Ville distribution, the rotated Loève transform, $D(t_0, f)$, is defined to be an expected value.

Stated in another way, the Wigner–Ville distribution is the *instantaneous estimate* of $D(t_0, f)$, representing the *dynamic spectrum* of the nonstationary signal $x(t)$ and therefore simpler to compute. Taking the complex conjugate of (3.9) readily shows that $D(t_0, f)$ is real, and, being the Fourier transform of a covariance function, it must be non-negative definite (see reference 14). In many ways, current discussions about positivity of the Wigner–Ville and similar distributions are reminiscent of those occurring in papers in the 1945–1970 era on whether the normalization $1/N$ or $1/(N - \tau)$ was "correct" for estimating lag-τ autocorrelations. The correct answer was that direct calculation of sample autocorrelations was a bad idea in any case and, given that the wrong estimate was being computed, the normalization was more or less irrelevant! The idea that the simple lagged correlations are "wrong" may sound peculiar, but there are several reasons why they are:

1. It is known [15] that, at least in some simple cases, multi-taper estimates of the spectrum are maximum likelihood. Thus the inverse Fourier transform of the multi-taper spectrum should also be the maximum likelihood estimate of autocovariance.

2. McWhorter and Scharf [16], extending previous work of Mullis and Scharf [17], also derive a multi-taper estimate of correlation using invariance arguments. Using a moving-average (MA) process as an example, they state the following:

   > *These curves indicate that the multiwindow estimates have better mean-squared error performance than the single window estimator in this scenario.*

   The term "multiwindow" used in this statement is another way of referring to "multi-taper."

3. Recently, Smith [18] showed that standard covariance estimates are biased.

3.3.1 Multi-taper Estimates

As discussed in Chapter 2, multi-taper estimates of a spectrum [19] are a class of estimates based on approximately solving the integral equation that expresses the projection of $dX(f)$ onto the Fourier transform of the data, $y(f)$. Taking the Fourier transform of the observed data—that is,

$$y(f) = \sum_{t=0}^{N-1} x(t) e^{-j2\pi ft} \tag{3.11}$$

and using the spectral representation (3.1) for $x(t)$—we have the *fundamental equation of spectrum estimation*:

$$y(f) = \int_{-1/2}^{1/2} K_N(f - \upsilon) dx(\upsilon) \tag{3.12}$$

where

$$K_N(f) = \frac{\sin N\pi f}{\sin \pi f} e^{-j2\pi f\left(\frac{N-1}{2}\right)} \tag{3.13}$$

is the *Dirichlet kernel.* There are several points that must be remembered about this fundamental equation:

1. Because we may take the inverse Fourier transform of $y(f)$ to recover $x(t)$ for $0 \le t \le N - 1$, $y(f)$ is a *sufficient statistic* and completely equivalent to the original data.
2. The *finite* Fourier transform $y(f)$ is *not* equivalent to the spectral generator $dX(\nu)$. Remember that $dX(\nu)$ is assumed to generate the entire data sequence for *all* t, not just the portion observed.
3. Despite definitions given in many elementary texts, $(1/N)|y(f)|^2$ is *not,* repeat *not,* the spectrum, even in the limit of large N. It is the periodogram, biased and inconsistent.
4. While (3.12) is formally a convolution of dX with a Dirichlet kernel, it is more constructive to think of this equation as a *Fredholm integral equation of the first kind.* As such, it does not have a unique solution. It does, however, have useful approximate solution. We mentioned above that "multi-taper

estimates" does not refer to a particular estimate, but rather to a class of estimates: The class is defined by the method used to form the necessarily approximate solution of the integral equation. Viewed in this way, spectrum estimation is, in reality, an *inverse problem* and therefore *ill-posed.*

Since multi-taper methods have been described in the book by Pereival and Walden [20], and in many papers that are becoming the "standard" in geophysics [21], elaborate description of their properties is unnecessary.[3]

3.3.2 Spectrum Estimation as an Inverse Problem

Recall the fundamental equation (3.12) and suppose we attempt to compute an eigensolution of the integral equation on the interval $(f - W, f + W)$ by assuming that the *observable* portion of dX has the expansion

$$d\hat{X}(f-\nu) = \sum_{k=0}^{k-1} x_k(f) V_k^*(\nu)\, d\nu \tag{3.14}$$

on the *local* frequency domain $(f - W, f + W)$. Here $V_k(\nu)$ is a *Slepian function* or discrete prolate spheroidal wave function; (see equation (2.13) of Chapter 2 for the definition of this function). Using the integral equation and properties of the Slepian functions, we obtain the raw expansion, or *eigencoefficients*

$$y_k(f) = \sum_{t=0}^{N-1} e^{-j2\pi f t} v_t^{(k)}(N, W) x(t) \tag{3.15}$$

as the Fourier transform of the data, $x(t)$, windowed by the kth Slepian sequence, $v_t^{(k)}(N, W)$. We retain the $K = 2NW$ coefficients corresponding to functions with eigenvalues $\lambda_k \approx 1$ for subsequent inference. *These eigencoefficients represent the information in the signal projected onto the local frequency domain.* This process resembles conventional, windowed spectrum estimation in that a fast Fourier transform (FFT) algorithm may be used for efficient computation, but differs in that standard estimates are best regarded as the first term of the multi-taper expansion.

Because the Slepian sequences are time-limited, they cannot be strictly band-limited, and the kth sequence has a fraction $1 - \lambda_k(N, W)$ outside the interval $(-W, W)$. Uncorrected, this out-of-band energy contributes bias, which can be severe for the higher-order eigencoefficients—that is, those of order $k \approx K$. Among the various ways of dealing with this exterior bias, the best method found to date is by *coherent sidelobe subtraction* as outlined in reference 28. Here we choose the bandwidth W to be large enough so that the bias on the lower-order terms is negligible and $x_k(f) \approx y_k(f)$ for $k << K$, thereby estimating the higher order eigencoefficients by $x_k(f) \approx y_k(f) - \hat{b}_k(f)$ for larger k. The bias estimate $\hat{b}_k(f)$ is formed from an exterior convolution of the Slepian sequence with an estimate of (3.14) and iterated. Denote the estimated eigencoefficients by $x_k(f)$ and collect them in the vector $\mathbf{x}(f)$:

[3] For extensions of the multi-taper method, see references 22 to 24. Two other extensions of the method are described in the books by Gubbins [25] and Weedon [26]. Another and particularly important extension is presented in reference 27.

$$\mathbf{x}(f) = [x_0(f), x_1(f), \ldots, x_{K-1}(f)]^T \tag{3.16}$$

To see the dependence on the bandwidth W of the estimate, recall that there are $K \approx \lfloor 2NW \rfloor$ windows with eigenvalues near 1. If the spectrum is flat *within the local domain*, then the coefficients are uncorrelated because the windows are orthogonal, and each window contributes two degrees of freedom, so estimates of the inner product $\mathbf{x}^H(f)\ \mathbf{x}(f)$ have $2K$ degrees of freedom, where the superscript H denotes Hermitian transposition. If W is too small, we have poor statistical stability, but if W is too large, the estimate has poor frequency resolution. Typically, W is chosen between $1.5/N$ and $20/N$ with a time–bandwidth product of 4 or 5 being a common starting point. Thus $W = 4/N$ or $5/N$, with the corresponding value $K = 6$ or 8 giving estimates with 12 or 16 degrees of freedom.

We must emphasize, however, that these only apply to the simplest forms of estimates and both quadratic inverse estimates (see references 29 and 30), and free parameter estimates of the type described therein give high-resolution estimates that are, within reason, largely independent of the choice of W. These estimates also give implicit extrapolations of the time series.

3.4 HIGH-RESOLUTION MULTI-TAPER SPECTROGRAMS

Beginning with the estimate of $d\hat{X}$ defined in (3.14), define the narrow-band process

$$X(t, f) = \int_{-W}^{W} e^{j2\pi t\xi} d\hat{X}(f \oplus \xi) \tag{3.17}$$

where $\oplus$ denotes *addition with the constraint* that $|\xi| < W$. On taking the inverse transform of the Slepian function, $X(t, f)$ becomes

$$X(t, f) = \sum_{k=0}^{K-1} \sqrt{\lambda_k}\, v_t^{(k)} x_k(f) \tag{3.18}$$

where λ_k is the kth eigenvalue. Clearly, the complex function $X(t, f)$ is not a time–frequency distribution but more akin to the output of a filter bank. Note, however, first, that if we write the approximate impulse response of the implied filters, they go from maximum phase at $t = 0$ through zero phase at $t = (N - 1)/2$, to minimum phase at $t = N - 1$. Second, $X(t, f)$ as defined here *extrapolates* the signal to t outside the interval $[0, N - 1]$ and thus resembles the Papoulis estimates [31]. The squared amplitude, $|X(t, f)|^2$, gives power as a function of time and frequency, as shown by

$$F(t, f) = \frac{1}{K} \left| \sum_{k=0}^{K-1} \lambda_k x_k(f)\, v_t^{(k)} \right|^2 \tag{3.19}$$

and it is an effective high-resolution spectrogram. Integrating this distribution over time yields the basic *multi-taper spectrum* estimate

$$S(f) = \frac{1}{k} \sum_{k=0}^{k-1} \lambda_k \, |x_k(f)|^2 \tag{3.20}$$

and thus gives a much more accurate distribution of power than time-frequency distributions that simply match $|y(f)|^2$. Similarly, integrating $F(t, f)$ over frequency gives

$$\int_{-\infty}^{\infty} F(t, f)\, df = \frac{1}{K} \sum_{n=0}^{N-1} \left| \sum_{k=0}^{K-1} \lambda_k \nu_t^{(k)} \nu_n^{(k)} \right|^2 |x(n)|^2 \tag{3.21}$$

or, approximately, the convolution of $|x(t)|^2$ with the sinc function $[\sin(2\pi Wt)/(\pi t)]^2$. Thus within a resolution interval $\Delta t = 1/(2W)$, power is approximately localized in time. Similarly, the narrow-band properties of the Slepian sequences imply that cross-terms are negligible for components separated in frequency by more than $2W$, so the time-frequency resolution area is of order 1. There are, however, two problems with this estimate: First, its distribution is "statistically unstable"; second, in common with time–frequency distributions of the form $x(t) \cdot y^*(f)$ (see Chapter 14 of Cohen [5]), this estimate can be thought of as

$$\int \gamma_L(\xi, f) e^{j2\pi\xi t} d\xi$$

and thus has "mixed" continuity properties caused by integrating across the expected δ-function sheets at 45° in only one of the two variables. A more serious criticism is that although such estimates satisfy enhanced marginal conditions, they appear to overlook the essential feature of correlation between frequencies more than $2W$ apart. Nonetheless, this "high-resolution" spectrogram represents, in applications where the spectrogram is useful, a vast improvement on the standard version. This estimate, like other multi-taper estimates, can obviously be extended to include overlapping data sections, so high-resolution spectrograms of long datasets can be formed by averaging the above estimates. Much more, indeed, is possible. Extending the definition (3.15) to make the base time b explicit

$$y_k(b, f) = \sum_{t=0}^{N-1} e^{-j2\pi ft} \nu_t^{(k)}(N, W)\, x(b+t) \tag{3.22}$$

we have, corresponding to (3.19),

$$F(b \oplus t, f) = \frac{1}{K} \left| \sum_{k=0}^{K-1} x_k(b, f) \nu_t^{(k)} \right|^2 \tag{3.23}$$

where $\oplus$ again represents a restricted sum, with $0 \le t \le N-1$. (Given the extrapolation properties of these estimates, mentioned above, this restriction is not strictly necessary, only conservative.) Section 3.4.1, on nonstationary quadratic-inverse estimates, discusses another way to improve on the standard spectrogram, and Section 3.4.2 shows how the basic expansion $X(t, f)$ may be used to estimate correlations between frequencies.

3.4.1 Nonstationary Quadratic-Inverse Theory

The problem of stability in the above estimate can be "solved" by *quadratic-inverse theory* [22, 28, 30, 32]. This is a way to generate minimum-variance unbiased estimates of second moment quantities directly from the eigencoefficients of the linear inverse solution without going through the ad-hoc procedure of generating the linear inverse, squaring, and then estimating the required second moments from these. Here we compute the eigensequences of the squared kernel (rigorously, the squared *truncated* kernel)

$$\alpha_l A_l(t) = N \sum_{m=0}^{N-1} \left[\frac{\sin 2\pi W (t-m)}{\pi (t-m)} \right]^2 A_l(m) \tag{3.24}$$

These sequences rapidly approach those of the continuous time problem [33], and there are approximately $4NW$ nonzero eigenvalues. Thus we have approximately $\alpha_l \sim 2NW - l/2$ for $l = 0, 1, \ldots, 4NW$, and because the variances of the quadratic-inverse coefficients are proportional to α_l^{-1}, the first few coefficients are nearly as stable as the standard multi-taper spectrum. The associated bases matrices

$$A_{jk}^{(l)} = \sqrt{\lambda_j \lambda_k} \sum_{t=0}^{N-1} v_t^{(j)} v_t^{(k)} A_l(t) \tag{3.25}$$

are real, symmetric, and trace-orthogonal; that is

$$\operatorname{tr}\left\{ \mathbf{A}^{(l)} \mathbf{A}^{(m)} \right\} = \alpha_l \delta_{lm} \tag{3.26}$$

The expansion coefficients corresponding to $F(t, f)$ are

$$\hat{p}_l(b, f) = \frac{1}{\alpha_l} \mathbf{X}^H(b, f) \mathbf{A}^{(l)} \mathbf{X}(b, f) \tag{3.27}$$

and so we have

$$P(t, f) = \sum_{l=0}^{N-1} \hat{p}_l(f) A_l(t) \tag{3.28}$$

The coefficients $\hat{p}_l(f)$ are often informative in their own right (see references 32 and 34). In particular, the zero-order function $A_0(t)$ is approximately constant, so $\mathbf{A}^{(0)} \approx \mathbf{I}$, where $\mathbf{I}$ is the identity matrix; and $\hat{p}_0(f)$ is approximately the standard multi-taper spectrum. The order 1 function $A_1(t)$ is approximately equal to $t - (N - 1)/2$. Thus $\mathbf{A}^{(1)}$ is zero on the diagonal and approximately constant on the sub- and superdiagonal, and $\hat{p}_1(f)$ is approximately the first *time derivative* of the spectrum, and so on. One useful ad-hoc quantity is $\hat{p}_1(f)/\hat{p}_0(f)$, approximating the time derivative of the natural logarithm $\ln S(t, f)$. For example, in reference 34, we find that $\hat{p}_1(f)$, computed from residuals of a global temperature series from 1854 to 1992, was almost uniformly negative across frequencies. While one must consider the series formally as nonstationary, the most reasonable explanation is not metaphysical, but simply that instrumentation and spatial coverage has improved since 1854. In this example, the quadratic inverse estimates are preferable to a spectrogram or the

Loève spectrum; the decrease in power is relatively small, and the data series has only 138 samples, so computing a spectrogram would be difficult and could be easily misinterpreted. Here, the negative derivative of the noise spectrum probably reflects little more than the improvements in instrumentation and spatial coverage that have occurred since 1854.

Expanding $F(t,f)$ of (3.19) in terms of the $A_l(t)$'s, it can be seen that the resulting coefficients, $F_l(f)$, are biased by α_l/K. However $F(t,f)$ is biased and positive, whereas truncation and the resulting Gibbs phenomenon can cause $P(t,f)$ to be negative.

Although spectrograms are insensitive to correlations between widely different frequencies, when the temporal evolution of the spectrum is slow, spectrograms form a useful intermediate class of time–frequency distributions. Quadratic-inverse estimates improve on the spectrogram by allowing for (a) changing power within the block, and (b) tests between blocks.

While the basic theory of multi-taper methods is usually written in terms of a finite block size N, we can obviously apply the same methods to overlapping time blocks to form spectrograms [29]. On each block, we estimate a dynamic spectrum $D(t, f)$, its frequency derivative,

$$D'(t, f) = \frac{\partial D(t, f)}{\partial f}$$

its time derivative,

$$\dot{D}(t, f) = \frac{\partial D(t, f)}{\partial t}$$

and perhaps higher order derivations.

These low-order terms are very stable with variances proportional to $1/\alpha_l$. Because $\hat{p}_0(f)$ is approximately the spectrum, $\hat{p}_1(f)$ is approximately the first time derivative of the spectrum, and so on. We can either make a "smoother" that uses these or, better, given $D(t, f)$ and its time derivative $\dot{D}(t, f)$, test if an estimate $D(t + \Delta, f)$ is "reasonable." Also, the "Nyquist sampling rate" for $F(t, f)$ is simply $\Delta = 1/(2W)$, so K samples spaced N/K are obtained in each time block. Thus, if the blocks are offset by Δ, then we have K estimates at each point of the time–frequency plane, so that averages and variances can be computed. The covariances between blocks can be computed, and tests for homogeneity of correlated variances are known, so the procedure can be used to test whether a choice of N and W is reasonable. Assuming that the estimate is reasonable, note that the average of the $F(t, f)$'s at each resampling time will be reasonably stable. Because of correlations between blocks, the stability of an average will be much less than $2K$ degrees of freedom, but the long lower tails characteristic of $\log\chi_2^2$ distributions are considerably suppressed. We use log spectra because (a) formally, the information content of a signal is measured by its Wiener entropy—a logarithmic measure, and (b) pragmatically, most engineering applications are designed for human use, and both the eye and ear have a logarithmic response. With the exception of helioseismology, it is difficult to find plots of power spectra which are not on a logarithmic (or decibel) scale. Thus we have a spectrogram with both good stability and time resolution!

3.4.2 Multi-taper Estimates of the Loève Spectrum

Taking the complex demodulates at two different frequencies, f_1 and f_2, an obvious estimate of their covariance is

$$\hat{\gamma}(f_1, f_2) = \frac{1}{K} \sum_{t=0}^{N-1} X(t, f_1) X^*(t, f_2) \tag{3.29}$$

where the normalization is proportional to the number of independent samples. Invoking orthogonality of the Slepian sequences, we may rewrite (3.29) in the equivalent form

$$\hat{\gamma}(f_1, f_2) = \frac{1}{K} \sum_{k=0}^{K-1} x_k(f_1) x_k^*(f_2) \tag{3.30}$$

This is the estimate given in reference 19; generally speaking, it works well (see references 35, 36, or 37 for more detail). An alternative motivation is that if we consider the product of two estimates of the approximate form

$$dX(f \oplus \xi) \sim \sum_{k=0}^{K-1} \hat{x}_k(f) V_k(\xi) d\xi \tag{3.31}$$

for $|\xi| < W$, then, guided by the continuity arguments of Section 3.3, we may use a weight $W(\xi_1, \xi_2) = \delta(\xi_1 - \xi_2)$, so that smoothing over a bandwidth W is done on the stationary frequency, no smoothing on the nonstationary frequency, and the same estimate is obtained. A similar smoothing scheme was proposed in reference 38 and applied effectively in reference 39.

Define the *dual-frequency coherence*

$$C(f_1, f_2) = \frac{\hat{\gamma}(f_1, f_2)}{[S(f_1) S(f_2)]^{1/2}} \tag{3.32}$$

We may then plot a dual-frequency spectrum, consisting of magnitude-squared coherence $|C(f_1, f_2)|^2$ and the associated phase $\arg\{C(f_1, f_2)\}$. Significance-level calculations for this *magnitude-squared coherence* (MSC) are exactly the same as they are for ordinary MSC calculations (see reference 40).

There are far too many extensions of this approach to describe in detail here; however, an indication of some directions should be mentioned:

(a) The correlation estimate in (3.30) can be extended to include a time delay—that is, ave$\{X(t, f_1)(X^*(t + \tau, f_2))\}$, which results in the quadratic form

$$\hat{\gamma}(f_1, f_2, \tau) = \sum_{j=0}^{N-1} \sum_{k=0}^{N-1} x_j(f_1) x_k^*(f_2) B_{jk}(\tau) \tag{3.33}$$

where B_{jk} is a function of the dely τ.

(b) We may use a similar quadratic form with, for example, $\mathbf{A}^{(1)}$, to test for energy transfer between frequencies. More generally, for a specific spectral pattern of interest, a weight $W(\xi_1, \xi_2)$ is chosen to emphasize it, and the integration over $-W < \xi_1, \xi_2 < W$ results in an appropriate weight matrix.

(c) We may treat $X(t, f)$ as a matrix, possibly scaling by $S(f)^{1/2}$, compute its singular-value decomposition (SVD), and then treat the dominant time-eigenvectors as a new time series.

(d) The same procedures can be applied to multivariate time series; in bivariate problems, compute ave$\{X(t, f_1)Y^*(t, f_2)\}$ or similar series can be "stacked" in the SVD process.

(e) In communications signals, it is common to encounter the same signal with sidebands reversed. In this case, the appropriate smoother would be perpendicular to the standard one and, as in reference 19, can be obtained by leaving the second coefficient unconjugated. Similar problems are encountered in dealing with other complex data [41].

3.5 SPECTRUM ANALYSIS OF RADAR SIGNALS

We now apply the theoretical ideas covered in this chapter to radar datasets representing three different environmental conditions: sea clutter on its own, weak target signal in clutter, and strong target signal in sea clutter. The target signal was due to the echo from a small piece of ice floating in the ocean under the dynamics of the ocean waves.[4] Note that these are actual data, not simulations, and consequently the clutter components in all three series are necessarily different and the energy in the clutter component of the series varies with enviromental conditions. Each series consists of 256 complex samples taken at intervals $\Delta T = 1.0$ ms.

Figures 3.1a–3.1c show high-resolution spectrograms of the data computed by the method of Section 3.4; they average the results of 10 sections of 229 samples, each offset by 3 samples. A time–bandwidth product of 6 was used with $K = 10$ windows on each section. The bandwidth is thus $\pm 6/(229\Delta T) = \pm 26$ Hz. The section offset used here is smaller than that recommended above, and the section averaging was used to suppress the lower tails of the $\log \chi^2$ distribution. (Normally, we would not attempt to compute a spectrogram from a sample of size 256.) The spectrogram of the clutter (Fig. 3.1a) shows a band near −110 Hz with more power than elsewhere, but otherwise the spectrogram is reasonably flat over the clutter spectrum. The contribution due to receiver noise is about 20 dB below the clutter spectrum.

By contrast, the spectrogram of the weak target (Fig. 3.1b) shows a strong, frequency-independent vertical stripe (due to clipping of the time series) near $t = 27$ ms as well as a second frequency stripe at about −25 Hz, in addition to the features visible in the clutter spectrogram.

With the strong target (Fig. 3.1c), the stripe centered near 0 Hz (representing the Doppler shift of the target signal) is much more obvious, the clutter band is still

[4] The radar data used in the spectral analysis reported herein were collected using the IPIX radar at a site on Cape Bonavista, Newfoundland, in late spring/early summer; the small piece of ice (commonly referred to as a "growler") was broken off an iceberg at a few kilometers from the coastline; the IPIX radar is described in Chapter 1.

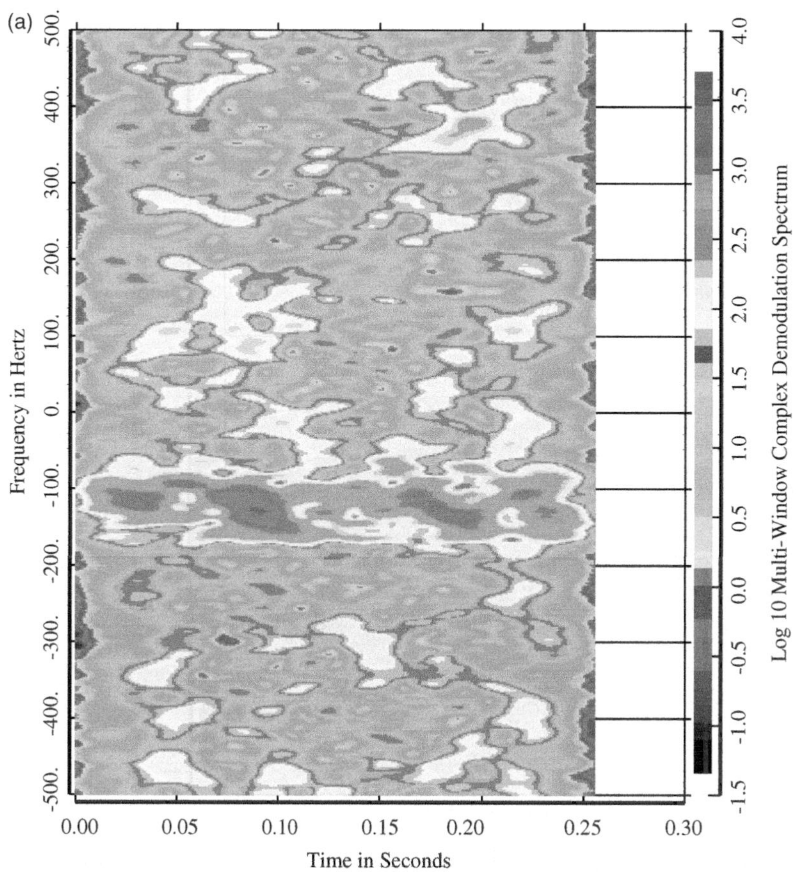

Figure 3.1 Dynamic spectra. (a) Radar clutter only data set, HH channel. (b) Radar weak target data, HH channel. (c) Radar strong target data, HH channel; part (a) of the figure is shown on this page, and parts (b) and (c) are shown on the next two pages.

there, and there is a weaker clutter image band, possibly due to a slight imbalance in the in-phase and quadrature channels of the coherent receiver.

Figures 3.2a–3.2c show the estimates of the corresponding Loève spectra, and the gain in information beyond that apparent in the spectrogram is striking. First, the diagonal bands evident in all three series show that the data are periodically correlated. In other words, sea clutter exhibits the property of *cyclostationarity*.[5]

[5] A stochastic process $x(t)$ is said to be *cyclostationary* in the wide sense if its second-order statistics (i.e., mean and autocorrelation function) exhibit *periodicity*, as in

- Mean: $\mu_x(t_1 + T) = \mu_x(t_1)$.
- Autocorrelation function: $R_x(t_1 + T, t_2 + T) = R_x(t_1, t_2)$ for all t_1 and t_2.

Modeling the process $x(t)$ as cyclostationary adds a new dimension—namely, *period T*—to the partial description of the process $x(t)$. A modulated process obtained by varying the amplitude, phase, or frequency of a sinusoidal carrier is an example of a cyclostationary process [42, 43].

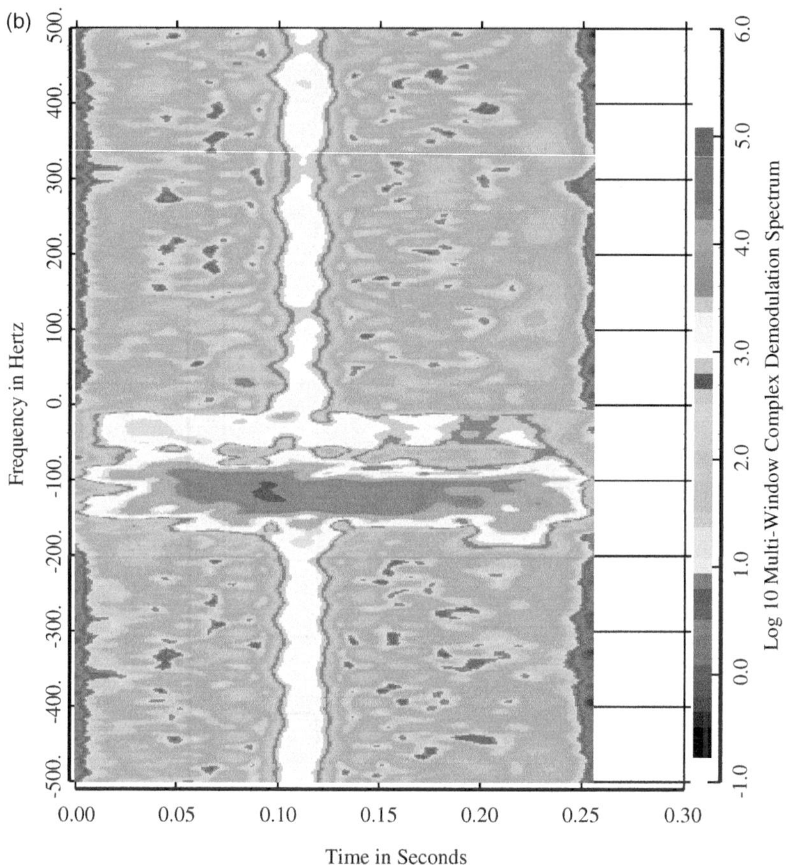

Figure 3.1 *(continued)*

This interesting property of sea clutter was *not* obvious in the spectrograms, nor expected. In Fig. 3.2a, it can be seen that the peak of the Loève spectrum for clutter is near −125 Hz, as before. In the weak target signal case, an extra peak centered at about −25 Hz can be seen in Fig. 3.2b. In the strong target signal case shown in Fig. 3.2c, the peak near zero is dominant, but, in contrast to the spectrogram, the periodic correlation of the clutter is still visible.

Figures 3.3a–3.3c present the corresponding Wigner–Ville distributions of the sea clutter on its own, weak target in sea clutter, and strong target in sea clutter, respectively. The important feature to observe here is the presence of a zebra-like pattern (alternating between dark and bright narrow stripes) in the images due to the presence of a target signal. This pattern occupies an area located between the instantaneous frequency plot of the target near 0 Hz and that of the clutter. This pattern is indeed a manifestation of the cross Wigner–Ville distribution terms due

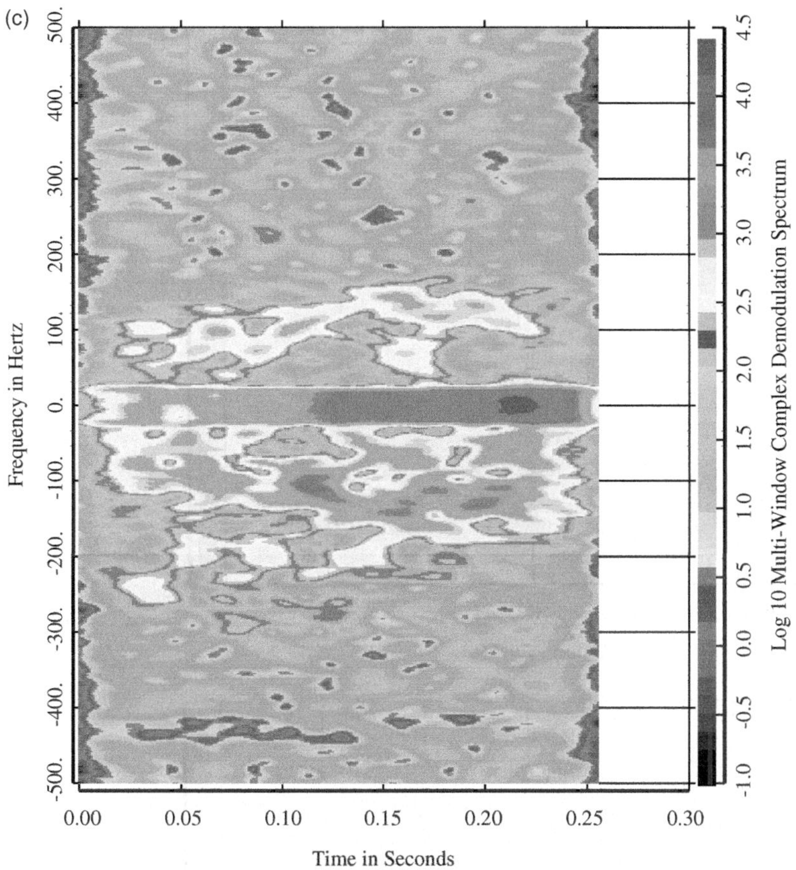

Figure 3.1 (*continued*)

to the combined presence of a target signal and clutter. Most importantly, the presence of this zebra-like pattern is found to be (1) fairly pronounced at relatively low target signal-to-clutter ratios, and (2) relatively robust to variations in the target signal-to-clutter ratio.

Although the high-resolution spectra estimates of Figs. 3.1 and 3.2 based on the method of multiple windows display the dynamic spectrum of the radar signals in ways that are similar (in some parts) and yet different (in other parts) from the corresponding Wigner–Ville distributions of Fig. 3.3, the important point to note from these two differently computed sets of images is that both approaches accentuate the differences between the different classes of radar signals in their own individual ways, making them more visible than in the original time series. Simply put, the power of these methods lies in their ability to make a weak target signal buried in a strong clutter background *visible* in signal-processing terms.

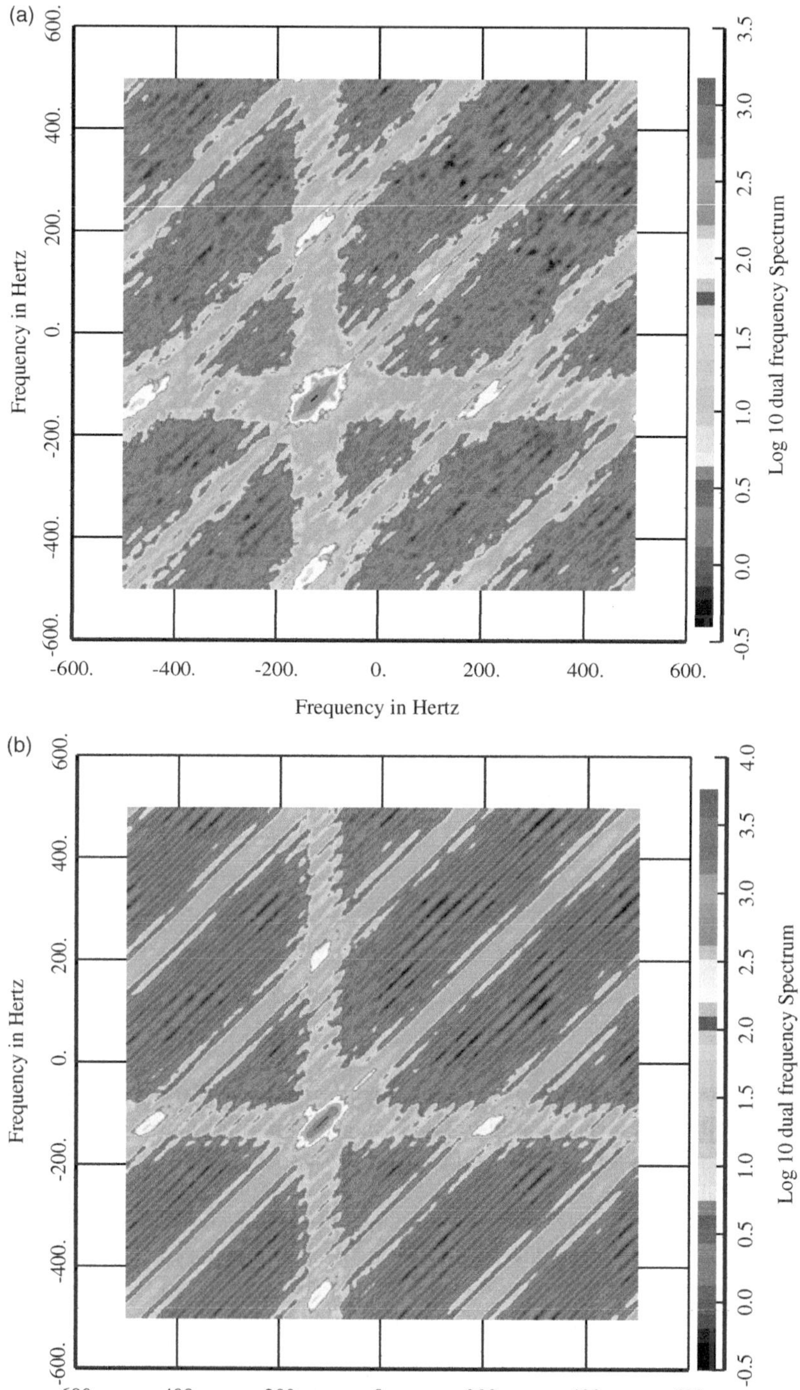

Figure 3.2 Dual-frequency plots. (a) Radar clutter only dataset, HH channel. (b) Radar weak growler data, HH channel. (c) Radar strong growler data, HH channel. Part (c) of the figure is shown on the next page.

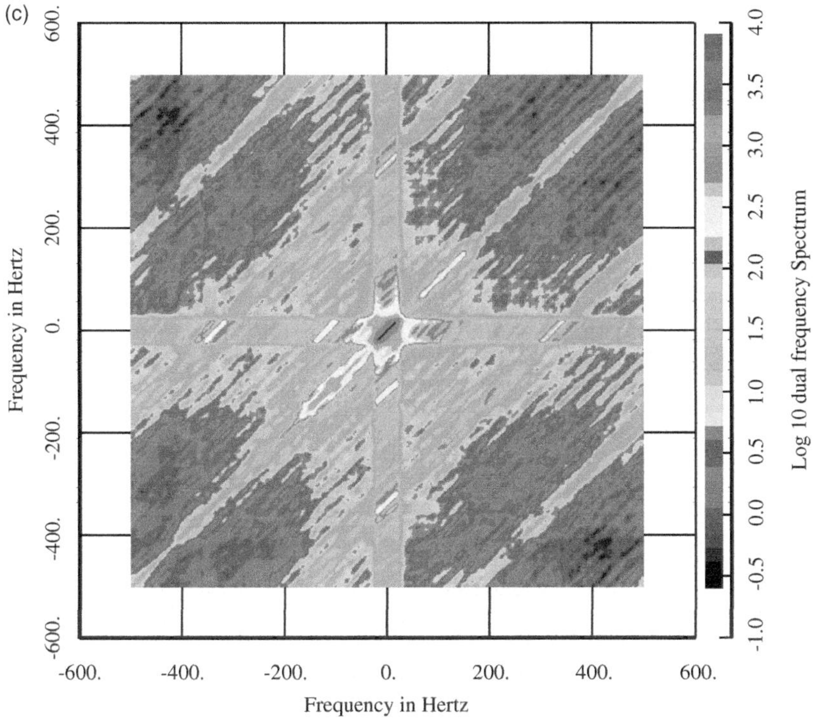

Figure 3.2 *(continued)*

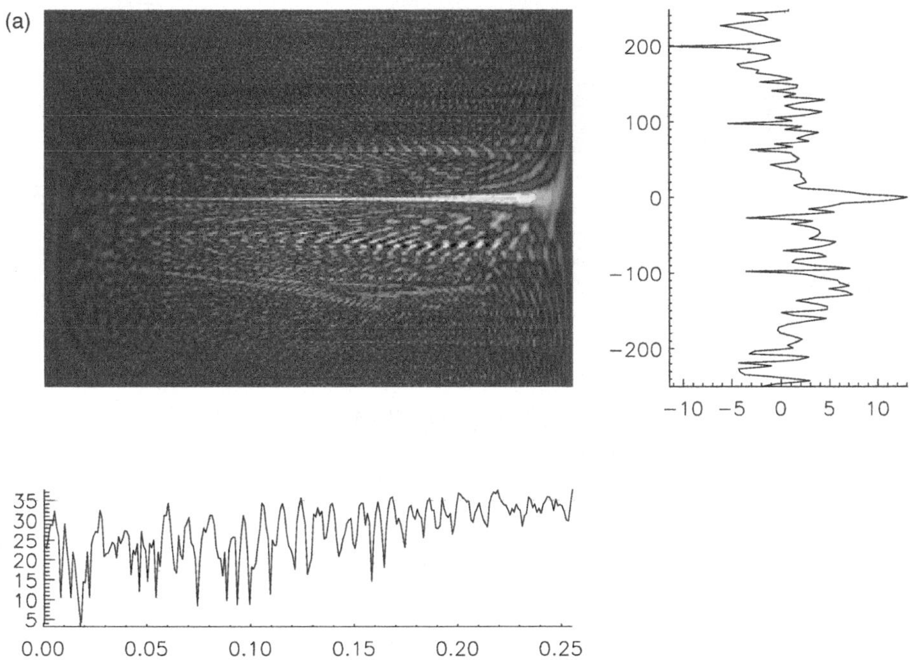

Figure 3.3 Wigner–Ville spectrum. (a) Clearly visible growler in sea clutter. (b) Barely visible growler in sea clutter. (c) Sea clutter alone. Parts (b) and (c) of Fig. 3.3 are shown on the next page.

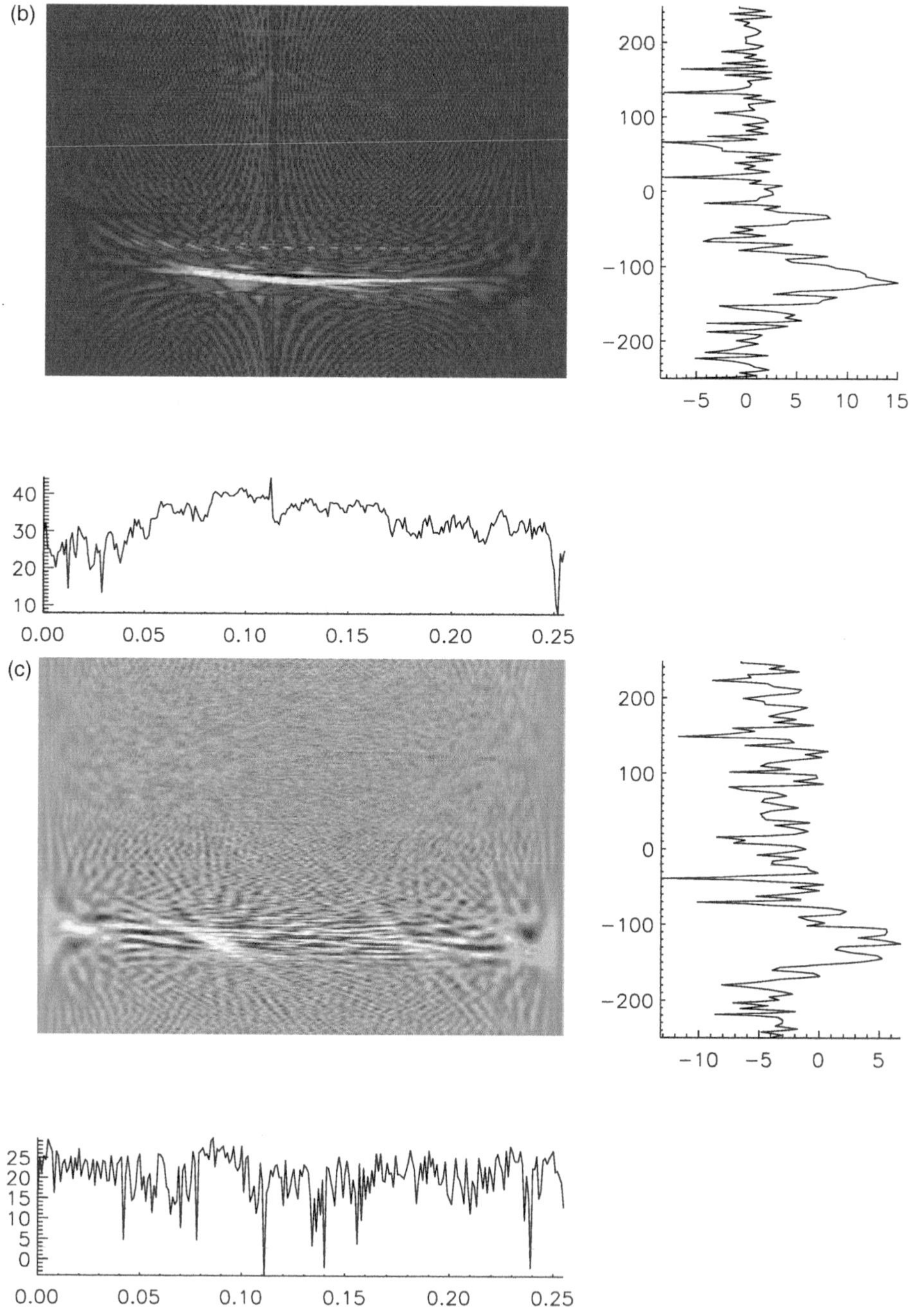

Figure 3.3 *(continued)*

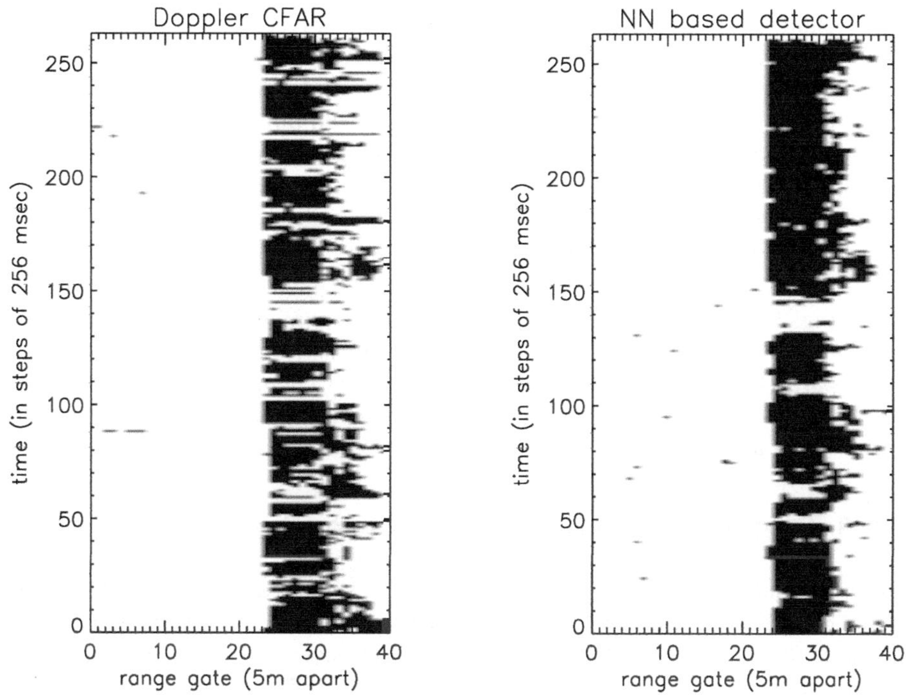

Figure 3.4 Postdetection results for the Doppler CFAR and NN receivers.

3.6 DISCUSSION

In this chapter, we did two important things:

(a) We presented a principal treatment of time–frequency analysis of nonstationary signals (albeit in a condensed manner) by invoking two mathematical tools: the Loève spectrum and the multi-taper method of spectrum analysis. Moreover, we showed that the well-known Wigner–Ville distribution is an instantaneous estimate of the dynamic spectrum, the latter being defined as a rotated version of the Loève spectrum.

(b) We applied the theory to investigate the time–frequency content of coherent radar returns from a sea surface, using real-life data. The investigation covered three entirely different environmental conditions:

(b.1) sea clutter acting alone

(b.2) weak (barely visible) target plus sea clutter

(b.3) strong target plus sea clutter

The results of the investigation may be summarized as follows:

- The time–frequency (spectral) and dual-frequency (coherence) images, respectively displayed in Figs. 3.1 and 3.2, delineate the underlying characteristics that differentiate the three environmental conditions; clutter by itself, clutter plus weak target, and clutter plus strong target. An important observation drawn from all three dual-frequency images in Fig. 3.2 is the unmistakable appearance of parallel strips along the direction (45°–225°). With sea clutter being the only component common to all three parts of Fig. 3.2, this observation suggests that sea clutter exhibits cyclostationarity. The implication of this property is that the underlying characterization of sea clutter embodies some form of modulation, which, indeed, will be confirmed by results presented in Chapters 4 and 5.

- The Wigner–Ville distribution (WVD) images pictured in the three parts of Fig. 3.3 also differentiate between the three environmental conditions mentioned above in a clear manner, which is distinctive of WVD. In particular, parts (b) and (c) of the figure exhibit a *zebra-like pattern*, which is attributed to the cross-WVD terms that arise due to the combined presence of a target signal and sea clutter.

3.6.1 Target Detection Rooted in Learning

We, as human beings, are "visual" thinkers. Given the two-dimensional images pictured in Figs. 3.1 through 3.3, it would seem feasible to build a *pattern classifier* that uses these images as "examples" of different environmental conditions. The postdetection results[6] of Fig. 3.4 demonstrate the feasibility of such an approach to target detection. Specifically, the figure displays the detected motion of a weak target in sea clutter, plotted versus the radar's range gate. The picture on the left-hand side of Fig. 3.4 displays the output of a conventional Doppler CFAR (constant false-alarm rate) receiver, while the picture on the right-hand side of the figure displays the corresponding output of a neural network (NN)-based receiver. The NN-based receiver consisted of a time-frequency analyzer based on the Wigner-Ville distribution, followed by two channels; one channel of the receiver was trained on sea clutter data alone, while the other channel was trained on sea clutter plus target data. The training in both cases were carried out using many examples collected under varying environmental conditions. Target detection in Fig. 3.4 is defined by the dark parts of the figure. With the weak target known to be moving along the ocean surface for the duration of time (approximately 65 seconds) represented by the figure, ideally the picture would consist of a dark vertical strip whose width is 5 meters (defined by the designed range gate of the radar receiver). Comparing the two parts of the figure, we may make two relevant observations:

1. The NN-based receiver exhibits a better detection performance than the conventional Doppler CFAR receiver.
2. The failure of the NN-based receiver to detect the weak target is attributed largely to the fact that occasionally the target would reside behind an ocean wave, which effectively leads to no radar returns from the target.

[6] The postdetection results presented in Fig. 3.4 are taken from the paper written by Haykin and Bhattacharya [44]. This paper also presents detailed description of the NN-based receiver used to compute the results of Fig. 3.4.

The CFAR processor is model-based, in that its design is based on a statistical model of the environment; in effect, prior knowledge about the underlying physics of the environment is embedded in the model. Accordingly, its performance is dependent on how good the model is. Nevertheless, the attractive feature of the CFAR processor is that it can be made *adaptive*, and therefore able to adapt to statistical variations in the environment. In contrast, the NN-based receiver (as conceived in reference 44) permits the real-life radar database (used to train the neural network) to "speak for itself", thereby by-passing the need for a statistical model. However, for this receiver to operate successfully under all possible environmental conditions, the radar database would have to be fully representative of the environment. This requirement, in turn, raises two practical issues:

- The physical effort and expense involved in collecting the radar database, which can be enormous if every conceivable occurrence of environmental behavior is to be covered.
- The time and effort expended in training the neural network, which can also be very demanding.

To get around the difficulty of having to work with a fully representative database, we may consider a *two-stage design strategy*: the first stage relying on *long-term memory*, and the second stage relying on *short-term memory*. For the first stage, a partially representative database covering "coarse behavior" of the environment may suffice. This database is used to train a supervised neural network; once the training is finished, the adjustable weights (parameters) of the neural network are fixed. Thus, knowledge contained in the training data is stored in the weights, and the trained neural network takes on the role of long-term memory. There are several powerful procedures for the supervised training of neural networks [45]; we may therefore say that the design of the first stage of the receiver is relatively straightforward. However, it is the design of the second stage representing short-term memory that is challenging, for which we are yet to devise a principled procedure.

This two-stage approach for designing a radar receiver is intuitively satisfying. In a loose sense, it mimics how the human brain performs its own target detection: The long-term memory represents knowledge (experience) gained by the brain through previous interactions with the environment. The short-term memory accounts for new knowledge (i.e., *innovation*) gained from current interactions with the environment. Guided by this motivation, building a two-stage radar receiver along the lines just described for target detection deserves serious consideration.

REFERENCES

1. M. Loève (1946). Fonctions aleatoires du second ordre, *Rev. Sci. Paris*, **84**, 195–206.
2. M. Loève (1963). *Probability Theory*, Van Nostrand, New York.
3. W. Koenig, H. K. Dunn, and L. Y. Lacy (1946). The sound spectrograph. *J. Acoustical Soc. Amer.*, **18**, 19–49.
4. J. C. Steinberg and N. R. French (1946). The portrayal of visible speech. *J. Acoustical Soc. Amer.*, **18**, 4–18.
5. L. Cohen (1995). *Time-Frequency Analysis*, Prentice-Hall, Englewood Cliffs, NJ.

6. FRANK. L. VERNON III (1989). *Analysis of Data Recorded on the ANZA Seismic Network*, Ph.D. thesis, University California, San Diego, 1989.
7. L. RAYLEIGH (1889). On the character of the complete radiation at a given temperature. *Philosophical Magazine*, XXVII, 460–469. (in *Scientific Papers by Lord Rayleigh*, Volume V, Article 160, 268–276, Dover Publications, New York, 1964).
8. L. RAYLEIGH (1903). On the spectrum of an irregular disturbance. *Philosophical Magazine*, **41**, 238–243. (in *Scientific Papers by Lord Rayleigh*, Volume V, Article 285, 98–102, Dover Publications, New York, 1964).
9. L. RAYLEIGH (1912). Remarks concerning Fourier's theorem as applied to physical problems. *Philosophical Magazine*, XXIV, 864–869. (in *Scientific Papers by Lord Rayleigh*, Volume VI, Article 369, 131–135, Dover Publication, New York, 1964).
10. R. A. FISHER (1922). On the mathematical foundations of theoretical statistics. *Phil. Trans. Roy. Soc., London, Series A*, **111**, pp. 309–368.
11. F. HLAWATSCH (1992). Regularity and unitary of bilinear time-frequency signal representations, *IEEE Trans. Information Theory* **38**, 82–94.
12. G. A. PRIETO, F. L.VERNON, G. MASTERS, AND D. J. THOMSON (2005). Multitaper Wigner-Ville spectrum for detecting dispersive signals from earthquake records. In *Proc. of the Thirty-Ninth Asilomar Conf. on Signals, Systems, and Computers*, 938–941.
13. W. MARTIN (1982). Time-frequency analysis of random signals, *Proc. ICASSP*, pp. 1325–1328.
14. P. FLANDRIN (1986). On the positivity of the Wigner–Ville Spectrum, *Signal Processing* **11**, 187–189.
15. P. STOICA AND T. SUNDLIN (1999). On nonparametric spectral estimation. *Circuits Systems Signal Process.*, **18**, 169–181.
16. L. T. MCWHORTER AND L. L. SCHARF (1998). Multiwindow estimators of correlation. *IEEE Trans. on Signal Processing*, **46**, 440–448.
17. C. T. MULLIS AND L. L. SCHARF (1991). Quadratic estimators of the power spectrum. In S. Haykin, editor, *Advances in Spectrum Analysis and Array Processing*, **2**, 1–57. Prentice-Hall.
18. S. T. SMITH (2005). Statistical resolution limits and the complexified Cramér-Rao bound. *IEEE Trans. on Signal Processing*, **53**, 1597–1609.
19. D. J. THOMSON (1982). Spectrum estimation and harmonic analysis, *Proc. IEEE* **70**, 1055–1096.
20. D. B. PERCIVAL AND A. T. WALDEN (1993). *Spectral Analysis for Physical Applications; Multitaper and Conventional Univariate Techniques*, Cambridge University Press, New York.
21. L. I. TAUXE (1993). Sedimentary records of relative paleointensity of the geomagnetic field; theory and practice, *Rev. Geophys* **31**, 319–354.
22. D. J. THOMSON (2000). Multitaper analysis of nonstationary and nonlinear time series data. In W. Fitzgerald, R. Smith, A. Walden, and P. Young, editors, *Nonlinear and Nonstationary Signal Processing*, 317–394. Cambridge University Press.
23. D. J. THOMSON, L. J. LANZEROTTI, AND C. G. MACLENNAN (2001). The interplanetary magnetic field: Statistical properties and discrete modes. *J. Geophys. Res.*, **106**, **15**,941–15,962.
24. D. J. THOMSON (2005). Quadratic-inverse expansion of the Rihaczek distribution. In *Proc. of the Thirty-Ninth Asilomar Conf. on Signals, Systems and Computers*, 912–915.
25. D. GUBBINS (2004). *Time Series Analysis and Inverse Theory for Geophysicists*. Cambridge University Press.
26. G. WEEDON (2003). *Time-Series Analysis and Cyclostatigraphy*. Cambridge University Press.
27. D. J. THOMSON AND A. D. CHAVE (1991). Jackknifed error estimates for spectra, coherences, and transfer functions. In S. Haykin, editor, *Advances in Spectrum Analysis and Array Processing*, volume 1, chapter 2, 58–113. Prentice-Hall.
28. D. J. THOMSON (1992). Quadratic-inverse spectrum estimates; applications to paleoclimatology, *Phil. Trans. R. Soc. Lond. A*. **332**, 539–597.
29. D. J. THOMSON (1990). Time series analysis of Holocene climate data, *Phil. Trans. R. Soc. Lond. A*. **330**, 601–616.
30. D. J. THOMSON (1994). An overview of multiple-window and quadratic-inverse spectrum estimation methods, *Proc. ICASSP*. **6**, 185–194.

31. A. Papoulis A new algorithm in spectral analysis and band-limited extrapolation, *IEEE Trans. Circuits Syst.* **CAS-22**, 735–742.
32. D. J. Thomson (1990). Nonstationary fluctuations in stationary time-series, *Proc. SPIE.* **2027**, 236–244.
33. F. Gori and C. Palma (1975). On the eigenvalues of sinc^2 kernel, *J. Phys. A: Math. Gen.* **8**, 1709–1719.
34. D. J. Thomson (1977). Dependence of global temperatures on atmospheric CO_2 and solar irradiance, *Proc. Natl. Acad. Sci. USA.* **94**, 8370–8377.
35. R. J. Mellors, F. L. Vernon, and D. J. Thomson (1996). Detection of dispersive signals using multi-taper dual-frequency coherence, in *Proceedings of the 18th Seismic Research Symposium on Monitoring a Comprehensive Test Ban Treaty, Annapolis, Maryland*, pp. 745–753.
36. R. J. Mellors, F. L. Vernon, and D. J. Thomson (1998). Detection of dispersive signals using multitaper dual-frequency coherence, *Geophys. J. Int.*, **135**, 146–154.
37. R. Schild and D. J. Thomson (1997). *The Q0957+561 Time Delay, Quasar Structure, and Microlensing*, Astronomical Time Series, D. Maoz et al. (eds.), Kluwer Academic Publishers, Dordrecht, pp. 73–84.
38. H. L. Hurd (1988). Spectral coherence of nonstationary and transient stochastic processes, Proc. Fourth IEEE ASSP Workshop on Spectrum Estimation and Modeling, Minneapolis, MN, pp. 387–390.
39. N. L. Gerr and J. C. Allen (1994). The generalized spectrum and spectral coherence of a harmonizable time series, *Digital Signal Processing* **4**, 222–238.
40. G. C. Carter (ed.) (1993). *Coherence and Time Delay Estimation*, IEEE Press, New York.
41. C. N. K. Mooers (1973). A technique for the cross-spectrum analysis of pairs of complex-valued time series, with emphasis on properties of polarized components and rotational invariants. *Deep-Sea Res.*, **20**, 1129–1141. (See comments and corrections by J. H. Middleton in vol. 29, pp. 1267–1269, 1982).
42. L. E. Franks (1969). *Signal Theory*, Prentice-Hall, Englewood Cliffs, NJ.
43. W. A. Gardner and L. E. Franks (1975). Characterization of cyclostationary random signal processes, *IEEE Trans. Information Theory* **IT-21**, 4–14.
44. S. Haykin and T. Bhattacharya (1997). Modular learning strategy for signal detection in nonstationary environment. In *IEEE Trans. Signal Processing*, **45**(6), 1619–1637.
45. S. Haykin (1999). *Neural Networks: A Comprehensive Foundation*, Prentice-Hall.

Part II

Dynamic Models

Chapter 4

Dynamics of Sea Clutter[†]

Simon Haykin, Rembrandt Bakker, and Brian Currie

4.1 INTRODUCTION

Nonlinear dynamics are basic to the characterization of many physical phenomena encountered in practice. Typically, we are given a time series of some observable(s), and the requirement is to uncover the underlying dynamics responsible for generating the time series. In a fundamental sense, the dynamics of a system are governed by a pair of nonlinear equations:

- A *recursive process equation*, which describes the evolution of the hidden state vector of the system with time:

$$\mathbf{x}_t = \mathbf{f}_t(\mathbf{x}_{t-1}, \mathbf{v}_{t-1}) \tag{4.1}$$

where the vector $\mathbf{x}_t$ is the state at discrete time t, $\mathbf{v}_{t-1}$ is the dynamic noise at time $t - 1$; the vector-valued function $\mathbf{f}$ is nonlinear and possibly a time-varying function (hence the subscript t).

- A *measurement equation*, which describes the dependence of observations (i.e., measurable variables) on the state:

$$\mathbf{y}_t = \mathbf{h}_t(\mathbf{x}_t, \mathbf{w}_t) \tag{4.2}$$

where the vector $\mathbf{y}_t$ is the *observation* at time t, and $\mathbf{w}_t$ is the measurement noise also at time t; the vector-valued function $\mathbf{h}$ is another nonlinear and possibly time varying function (hence the subscript t).

Equations (4.1) and (4.2) define the *state-space model* of a nonlinear, time-varying dynamic system in its most general form. The exact form of the model

[†] The material presented in this chapter is an expanded version of the following paper: Simon Haykin, Rembrandt Bakker, and Brian Currie (2002). Uncovering nonlinear dynamics: The case study of sea clutter, *Proceedings IEEE Special Issue on Applications of Nonlinear Dynamics* **90** (5), 860–881.

adopted in practice is influenced by two perspectives that are in a state of "tension" with each other:

- Mathematical tractability
- Physical considerations

Mathematical tractability is at its easiest when the system is *linear* and the dynamic noise $\mathbf{v}_t$ and measurement noise $\mathbf{w}(t)$ are both *additive and modeled as independent white Gaussian noise processes*. Under this special set of conditions, the solution to the problem of uncovering the underlying dynamics of the system is to be found in the celebrated *Kalman filter* [1]. In a very clever way, the Kalman filter solves the problem by exploiting the fact that there is a one-to-one correspondence between the given sequence of observable samples and the sequence of *innovations* derived from one-step predictions of the observables; the innovation is defined as the difference between the observation y_t and its minimum mean-square error prediction, given all previous values of the observation up to and including time $t-1$.

Unfortunately, many of the dynamic systems encountered in practice are *nonlinear*, which makes the problem of uncovering the underlying dynamics of the system a much more difficult proposition. Consider, for example, the time series displayed in Fig. 4.1. These time series, made up of sampled signal amplitude versus time, were obtained by the instrument-quality, multifunction (mechanically scanned) IPIX radar,[1] which was configured to monitor a patch of the ocean surface at a low grazing angle. The radar was mounted at a site in Dartmouth, Nova Scotia, on the East Coast of Canada, at a height of about 30 m above the sea level. The radar was operated in a *dwelling mode* so that the dynamics of the *sea clutter* (i.e., radar backscatter from the ocean surface) recorded by the radar would be entirely due to the motion of the ocean waves and the natural motion of the sea surface itself. Throughout the chapter we will make extensive use of three different datasets. Two datasets were measured at low wave-height conditions (0.8 m) and are labeled *L*1 and *L*2. For the third dataset, labeled *H*, the wave height was higher (1.8 m); the characteristics of these datasets are summarized in Appendix A.

From the viewpoint of dynamic systems characterized by the pair of equations (4.1) and (4.2), we may identify six potential sources responsible for the difficulty in understanding the complex appearance of the time series in Fig. 4.1:

1. The dimensionality of the state.
2. The function $\mathbf{f}$ governing the nonlinear evolution of the state with time.
3. The possible presence of dynamic noise complicating the evolution of the state with time.
4. The function $\mathbf{h}$ governing dependence of the radar observable on the state.
5. The unavoidable presence of measurement noise due to imperfections in the instruments used to record the sea clutter data.
6. The inherently nonstationary nature of sea clutter, hence the explicit dependence of both nonlinear functions in (4.1) and (4.2) on time t.

[1] The IPIX radar is described in Chapter 1.

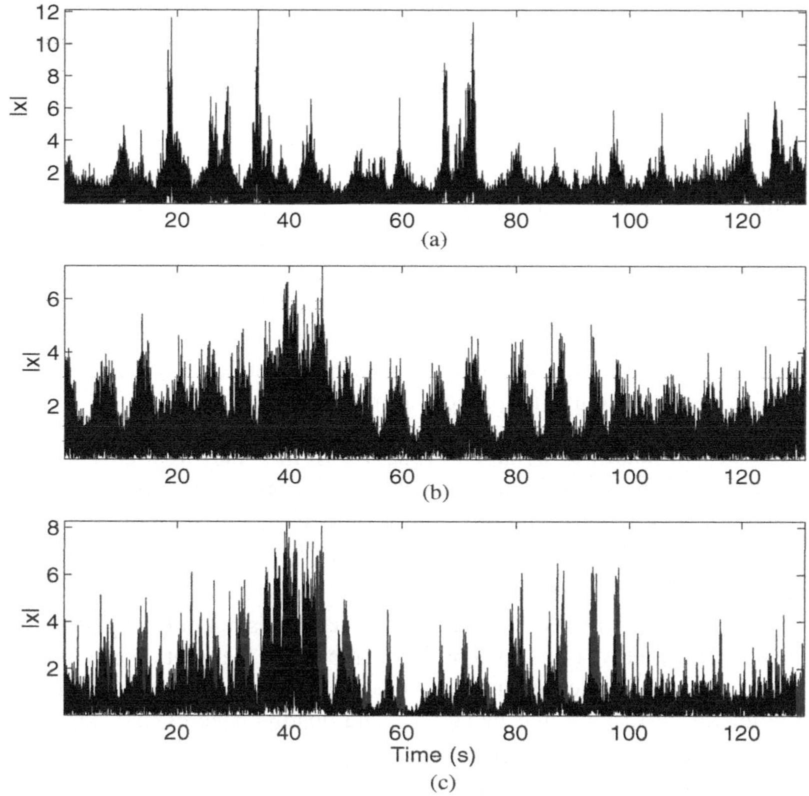

Figure 4.1 Radar return plots. (a) Dataset *L*2, VV polarization; (b) dataset *H*, VV polarization; and (c) dataset *H*, HH polarization. $|x|$ is the magnitude of the complex envelope of the return signal. The units of x are normalized.

Many, if not all, of these parameters/processes are unknown, which makes the uncovering of the underlying dynamics of sea clutter into a challenging task.

Random-looking time-series, such as those of Fig. 4.1, can be modeled at various levels of sophistication. The crudest form is to look at the *probability density function* (pdf) of the data, ignoring any type of correlation in time. At the next level, correlations in time are modeled by a linear or higher order relationship, and the residuals are described by their pdf. A third level of sophistication is sometimes possible for systems that exhibit *low-dimensional dynamics* [2–7]. For a subset of these systems, namely, *deterministic chaotic systems*, the time series can be described completely in terms of nonlinear evolutions, and, assuming a perfect model and noise-free measurements, there are no residuals at all. The deterministic chaos approach has enormous potential in that it makes it possible to reproduce the mechanism underlying the experimental data with a computer model. It has attracted the attention of numerous researchers in the natural and applied sciences, trying to identify if their data are close to being chaotic and lend themselves for a deterministic modeling approach. *Chaos theory* itself is motivated by earlier works of Kaplan and Yorke [8], Packard et al. [9], Takens [10], Mañé [11], Grassberger and Procaccia [12], Ruelle [13], Wolf et al. [14], Broomhead and King [15], Sauer et al. [16], Sidorowich [17], and Casdagli [18]. Indeed, it was these papers that aroused interest in deterministic chaos as a possible mechanism for explaining the underlying dynamics of sea clutter [19–23]. Unfortunately, for reasons that will be explained later in Section 4.3, currently available state-of-the-art algorithms used to estimate the chaotic invariants of sea clutter produce inconclusive results, which cast serious doubts on deterministic chaos as a possible mathematical basis for the nonlinear dynamics of sea clutter. This conclusion has been reinforced further by the inability to design a reliable algorithm for the dynamic reconstruction of sea clutter.

All along, our own primary research interests in sea clutter have been driven by the following issues of compelling practical importance:

- Sea clutter is a nonlinear dynamic process with *time* playing a critical role in its characterization. By contrast, much of the effort devoted to the characterization of sea clutter during the past 50 years has focused on the statistics of sea clutter, with little attention given to time [24–30], other than adapting to time-varying statistical parameters.
- Understanding the nonlinear dynamics of sea clutter is not only important in its own right, but it will have a significant impact on the joint detection and tracking of a point target on or near the sea surface. Such targets include a low-flying aircraft, small marine vessels, and floating hazards (e.g., ice).
- Identifying the particular part of the expanding literature on nonlinear dynamics, which is applicable to reliable characterization of sea clutter.

This chapter is written with these objectives in mind, given what we currently know about the statistics and dynamics of sea clutter.

The rest of the chapter is organized as follows. Section 4.2 presents a tutorial review of the classical models of sea clutter, with primary emphasis on the compound *K*-distribution. Section 4.3 presents a critical review of results reported in the literature on the application of deterministic chaos analysis to sea clutter. The discussion presented therein concludes that the discovery that a real-world *experimental* time series is chaotic has a high risk of being a *self-fulfilling prophecy*. We justify this statement by revisiting earlier claims that sea clutter is the result of a deterministic chaotic process. In Section 4.4, we go back to first principles in modulation theory and present new experimental results demonstrating that sea clutter is the result of a *hybrid continuous-wave modulation process that involves amplitude- as well as frequency-modulation*; this section also includes a time-varying, data-dependent autoregressive model for sea clutter, which, in a way, relates to our earlier work on the autoregressive modeling of radar clutter in an air-traffic control environment [31–34]. Section 4.5, the final section of the chapter, presents conclusions and an overview of the current directions of research on recursive-learning models that may be relevant to the nonlinear dynamics of sea clutter.

4.2 STATISTICAL NATURE OF SEA CLUTTER: CLASSICAL APPROACH

Sea clutter, referring to the radar backscatter from the sea surface, has a long history of being modeled as a stochastic process, which goes back to the early work of Goldstein [24]. One of the main reasons for this approach has been the random-looking behavior of the sea clutter waveform. In the classical view, going back to Boltzmann, the irregular behavior of a physical process encountered in nature is believed to be due to the interaction of a large number of degrees of freedom in the system, hence the justification for the statistical approach.

There are three signal domains of the radar waveform in which the clutter properties need to be characterized: amplitude, phase, and polarization. Noncoherent radars measure only the envelope (amplitude) of the clutter signal. Coherent radars are able to measure both signal amplitude and phase. Polarimetric effects are evident in both types of radar. Before discussing these effects, some background on the characterization of the sea surface and consideration of the geometry of a low-grazing-angle radar is desirable.

4.2.1 Background

It is the nature of the surface roughness that determines the properties of the radar echo [35]. The roughness of the sea surface is normally characterized in terms of two fundamental types of waves. The first type is termed *gravity waves*, with wavelengths ranging from a few hundred meters to a fraction of a meter; the dominant restoring force for these waves is the force of gravity. The second type is smaller *capillary waves* with wavelengths on the order of centimeters or less; the dominant restoring force for these waves is surface tension.

The gravity waves, which describe the macrostructure of the sea surface, can be further subdivided into sea and swell. Sea consists of wind waves, which are steep short-crested waves driven by the winds in their locale. Swell consists of waves of long wavelength, nearly sinusoidal in shape, produced by distant winds. The very irregular appearance of the sea surface is due to interference of the various wind and swell waves and to local atmospheric turbulence. Near coastlines, ocean currents (usually tidal currents) may cause a considerable increase in the wave heights due to their interference with wind and swell waves. The microstructure of the sea surface—consisting of the capillary waves—is usually caused by turbulent gusts of wind near the surface.

Waves are primarily characterized by their length, height and period. The phase speed is the ratio of wave length over wave period. Wave length and period (hence, phase speed) can be derived from the dispersion relation [36]. Wave height fluctuates considerably. A commonly reported measure is *significant wave height,* defined as the average peak-to-trough height of the one-third highest waves. It indicates the predominant wave height.

To provide a simple metric to indicate qualitatively the current sea conditions, the concept of *sea state* has been introduced in the literature. Table 2-1 in reference

35 links expected wave parameters, such as height and period, to environmental factors including wind speed, duration, and fetch. The frequently used short form gives the sea-state number only. So we speak of sea state 4, 5, or 6, for example, depending on how rough the sea surface is.

The angle at which the radar beam illuminates the surface is called the *grazing angle*, φ, measured with respect to the local horizontal. The smallest area, A_r, of the sea surface within which individual targets can no longer be individually resolved is termed the resolution cell, whose area is given by

$$A_r = R\theta_b \left(\frac{c\tau}{2}\right) \sec\varphi \tag{4.3}$$

where R is range, θ_b is the azimuthal beamwidth of the antenna, c is the speed of light, τ is the radar pulse length, and φ is the grazing angle.

The backscatter power (square of the amplitude) has been studied at two different time scales. Studies reported in reference 37 have produced empirical models relating to the long-term (over several minutes) average, given as the normalized radar cross section, to various parameters, including grazing angle, radar frequency and polarization, and wind and wave conditions.

Polarimetric Effects

One of the dominant scattering mechanisms at microwave frequencies and low-to-medium grazing angles is *Bragg scattering*. It is based on the principle that the returned signals from scatterers that are half a radar wavelength apart (measured along the line of sight from the radar) reinforce each other since they are in phase. At microwave frequencies, the Bragg scatter is from capillary waves. It has long been observed that there is a difference in the behavior of sea backscatter depending on the transmit polarization.[2] Horizontally polarized (HH) backscatter has a lower average power as compared to the vertically polarized (VV) backscatter, as predicted by the composite surface theory and Bragg scattering [38]. As a consequence, most marine radars operate with HH polarization. However, the HH signal often exhibits large target-like *spikes* in amplitude, with these spikes having decorrelation times on the order of one second or more.

Figure 4.2 shows the evolution of the Doppler spectrum versus time for the coherent data used to generate the amplitude plots of Fig. 4.1. In this case of incoming waves, the HH spectrum on average is shifted further from the frequency origin (i.e., has a higher mean Doppler frequency), and at the times of strong signal content, the HH spectrum may reach higher frequencies than does the VV spectrum.

The differences in the spectra suggest that different scatterers are contributing to the HH and VV returns. They can be partially explained from conditions associated with *breaking waves*. The breaking waves contribute to the bunching of scatterers, consistent with arguments for the applicability of the compound K-distribution

[2] A *signal's polarization* is designated by a two-letter combination TR, where T is the transmitted polarization (H or V) and R is the received polarization (H or V); thus we speak of four kinds of polarization: HH, HV, VH, and VV.

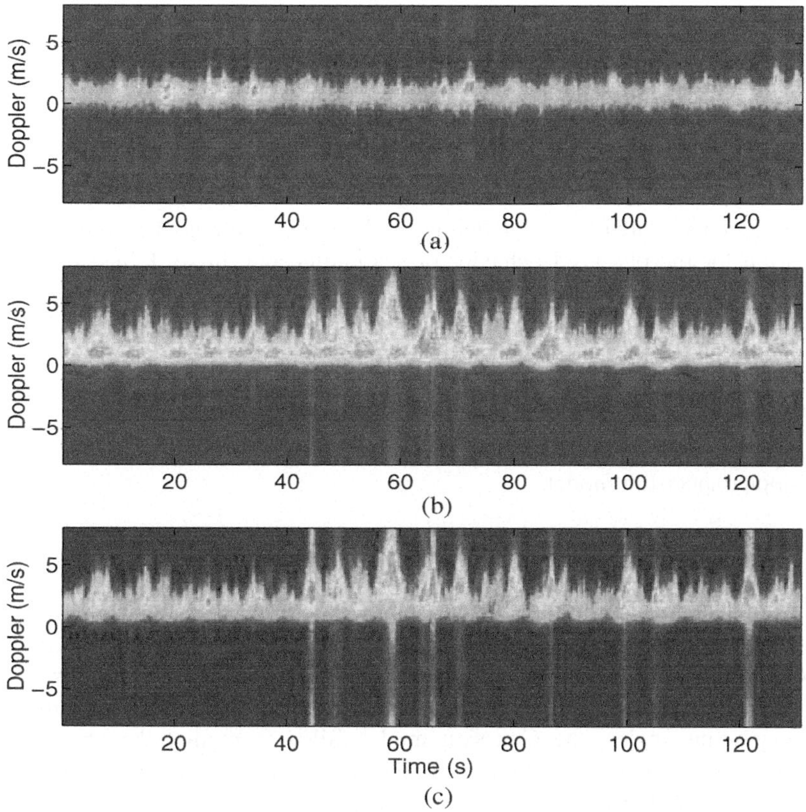

Figure 4.2 Time–Doppler plots. (a) Dataset *L*2, VV polarization; (b) dataset *H*, VV polarization; and (c) dataset *H*, HH polarization, using a window size of 0.5 s.

[39]. With the scatterers bunched at or near the crest of the breaking wave, there is the opportunity for a multipath reflection from the sea surface in front of the wave. The polarization dependence arises from the relative phase of the direct and surface-reflected paths. For VV polarization, the *Brewster effect* may lead to strong cancellation of the return, whereas the HH polarization will exhibit a strong (possibly spiky) return [39]. The Brewster angle is the particular angle of incidence for which there is no reflected wave when the incident wave is vertically polarized.

From *X*-band scatterometer data from advancing waves, Lee et al. [40] identified VV-dominant, comparatively short-lived "slow (velocity) scatterers" and HH-dominant longer-lived "fast (velocity) scatterers." Because the water particles that define a breaking wave crest necessarily exceed the orbital acceleration of the linear-wave group that initiates the nonlinear evolution of the wave structure, the fact that fast scatterers are observed is not surprising. Sea spikes from advancing waves are collocated with the fastest scatterers, which are identified with the wave crest. Based on experimental data for approaching waves, Rino and Ngo [39] suggest that the VV backscatter is responding to slower scatterers confined to the back-side of the wave while HH is responding to the fast scatterers near the wave crest. The HH response to the backside scatterers (presumed to be Bragg-like structures) may be suppressed due to the angular dependence of the Bragg scattering.

4.2.2 Current Models

There are two goals related to the modeling of clutter. The first goal is to develop an explanation for the observed behavior of sea clutter and, in so doing, to gain insight into the physical and electromagnetic factors that play a role in forming the clutter signal. Based on the success of the first goal, the second goal is to produce a model, ideally physically based, with which a representative clutter signal can be generated, to extend receiver algorithm testing into clutter conditions for which sufficient real data are unavailable. Two current models that seek to address the second goal (at least in one of the signal domains) are the compound *K*-distribution model and the Doppler spectrum model.

Compound K-Distribution

Characterization of the amplitude fluctuations of the sea backscatter signal is a continuing source of study. Much of the early work in fitting amplitude distributions was based on the use of a Gaussian model, implying Rayleigh distributed amplitudes. However, it was soon found that operating with increased radar resolution and at low grazing angles, the Gaussian model failed to predict the observed increased occurrence of higher amplitudes. Thereafter, researchers began using two-parameter distributions to empirically fit these longer tails. Such distributions include Weibull [25], lognormal [41], and K [29, 30]. Use of the latter has led to the development of the compound *K*-distribution.

The nature of the sea surface, with its two fundamental types of waves—short capillary and wind waves, and longer gravity waves—suggests the utility of a model composed of two (or perhaps more) components. This approach, in various forms, has been proposed by several researchers (e.g., references 28 and 42). One such approach is the *compound K-distribution* [28, 29]. From experimental studies, it was found that over short periods, on the order of a few hundred milliseconds, the sea clutter amplitude can be fitted reasonably well with a Rayleigh distribution. Then, averaging the data over periods on the order of 30 ms to remove the fast fluctuation, the resulting longer-term variation could be fitted with a Chi (or root-gamma) distribution. The proposed model is one in which the overall clutter amplitude is modeled as the product of a Rayleigh-distributed term and a root-gamma distributed term. The overall amplitude distribution $p(x)$ is given by

$$p(x) = \int_0^\infty p(x|y)\,p(y)\,dy \tag{4.4}$$

where $p(x|y)$ is the conditional pdf of x given y (both assumed to be scalar here), and $p(y)$ is the marginal pdf of y. The two pdfs within the integral are respectively defined by

$$p(x|y) = \frac{\pi x}{2y^2}\exp\left(-\frac{\pi x^2}{4y^2}\right), \qquad 0 \le x \le \infty \tag{4.5}$$

and

$$p(y) = \frac{2b^{2v}}{\Gamma(v)} y^{2v-1}\exp(-b^2y^2), \qquad 0 \le y \le \infty \tag{4.6}$$

where $\Gamma(0)$ is the gamma function. Equation (4.5) shows $p(x|y)$ to be Rayleigh distributed, with the mean level determined by the value of y. The distribution of y given by (4.6) is Chi or root-gamma. Substituting (4.5) and (4.6) into (4.4) yields

$$p(x) = \frac{4c}{\Gamma(v)}(cx)^v K_{v-1}(2cx), \qquad 0 \le x \le \infty \tag{4.7}$$

where $c = b\sqrt{\pi/4}$, and $K_{\nu-1}(0)$ is the *modified Bessel function of the third kind of order* $\nu - 1$, and ν is the *shape parameter*. The resulting overall distribution given by (4.7) is the K-distribution; hence, the model is termed the compound K-distribution model. The Rayleigh-distributed component may be considered as modeling the short-term fluctuation of the scatterers, while the root-gamma distributed component represents the modulation of the intensity of the scattering in response to the gravity waves. Since sea clutter is locally Rayleigh distributed (resulting from application of the central limit theorem within a patch), it appears that the non-Rayleigh nature of the overall clutter amplitude distribution is due to bunching of the scatterers by the sea wave structure, rather than being due to a small number of effective scatterers [29].

For detailed statistical characterization of sea clutter, we also need to consider the correlation properties of the clutter amplitude. Figure 4.3 shows a typical plot of the autocorrelation of the VV signal on two time scales. The left trace, based on

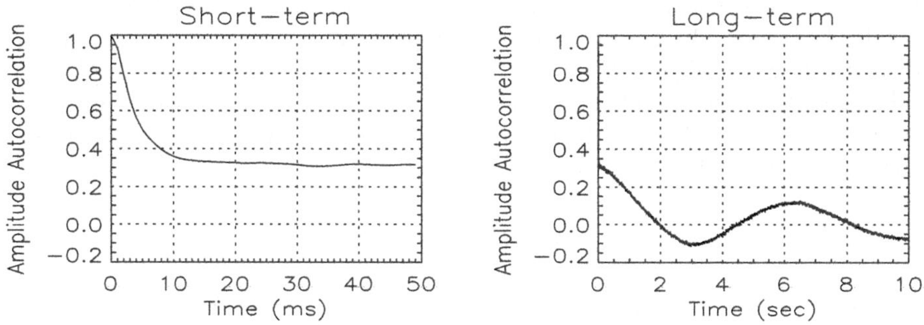

Figure 4.3 Plots showing the two time scales of the clutter amplitude autocorrelation, for the data of Fig. 4.1b. The left graph shows the quick initial decorrelation, on the order of a few milliseconds, of the fast-fluctuation component. The right graph shows the slowly decaying and periodic correlation of the slow-fluctuation component. The oscillation reflects the periodicity of the swell wave.

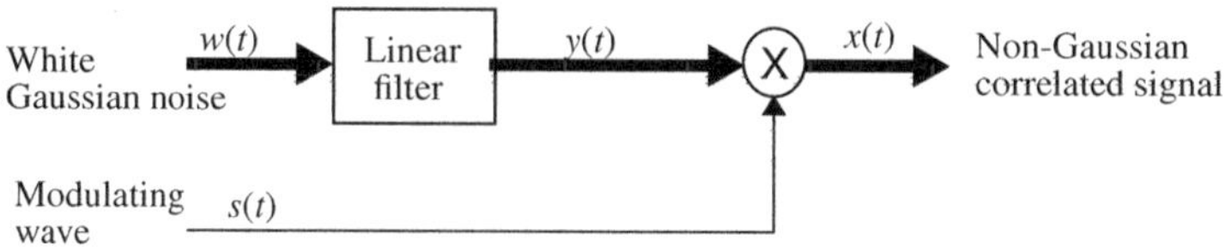

Figure 4.4 Generic model for generating complex non-Gaussian correlated data. The thick line denotes the flow of complex quantities. (After reference 43.)

a sample period of 1 ms, shows that the correlation due to the fast fluctuation component is under 10 ms. The right trace shows the long-term correlations, on the order of 1 s. Note, however, the apparent periodicity of the long-term autocorrelation, on the order of 6.5 s. This oscillation reflects the periodicity of the swell wave.

For generating *K*-distributed clutter, both Ward et al. [29] and Conte et al. [43] have suggested the same basic structure, shown in Fig. 4.4. Complex white Gaussian noise, $w(t)$, is passed through a linear filter, whose coefficients are chosen to introduce the desired short-term correlation. The output of the filter is still Gaussian distributed, so that the amplitude of $y(t)$ is Rayleigh distributed. The modulating term $s(t)$ is a real non-negative signal with a much longer decorrelation time compared to $y(t)$. To generate a *K*-distributed amplitude, the $s(t)$ should be drawn from a Chi distribution. Addressing the long-term correlation of the clutter requires generating correlated Chi-distributed variates $s(t)$. It is not possible to produce an arbitrary correlation, but some useful results have been reported. Gaussian variates are passed through a simple first-order autoregressive filter, then converted using a memoryless nonlinear transform into Chi variates with an exponentially decaying autocorrelation. Watts [44] parameterizes the form of the autocorrelation in terms of the clutter decorrelation time and the shape parameter of the *K*-distribution. Details can be found in Watts [44], Tough and Ward [45], and Conte et al. [43].

Doppler Spectrum

A coherent radar is able to measure both the amplitude and phase of the received signal. The received baseband signal is a complex voltage, given in terms of either (a) its in-phase (I) and quadrature (Q) components, or (b) its magnitude (amplitude) and phase angle. Movement of the scatterer relative to the radar causes a pulse-to-pulse change in the phase of the radar echo. This phase change is equivalent to a Doppler frequency shift, given as

$$f = \frac{2v}{\lambda} \tag{4.8}$$

where λ is the radar wavelength, and f is the Doppler frequency-shift resulting from the movement at a velocity v along the radial between the radar and the scatterer. The *Doppler spectrum* of sea clutter results from two main processes: The spread about the mean Doppler frequency is a manifestation of the random motion of the unresolved scatterers, while the displacement of the mean Doppler frequency maps the evolution of the resolved waves. Tracking the evolution of the Doppler spectrum versus time can provide insight into the scattering mechanisms and can identify properties that a sea clutter model should possess.

Note that in a realistic sea surface scenario, there will be a continuum of waves of various heights, lengths, and directions. This continuum is typically characterized by a wave-frequency spectrum (or wave-height spectrum), describing the distribution of wave height versus wave frequency. There are a number of models relating environmental parameters such as wind speed to the frequency spectrum [46]. The frequency spectrum can then be extended to the directional frequency spectrum by introducing a directional distribution [47]. Under the assumption of linearity, the combined effect is the superposition of all the waves, calculated by integrating across the appropriate range of directional wave numbers. In reality, the final surface is a nonlinear combination of the continuum of waves.

Walker [48] studied the development of the Doppler spectra for HH and VV polarizations as the breaking wave passed the radar sampling area. Coincident video images were taken of the physical wave. Three types of scattering regimes appear to be important: Bragg, whitecap, and spike events. Walker [49] proposes a three-component model for the Doppler spectrum based on these regimes.

1. *Bragg scattering*: This regime makes VV amplitude greater than HH. Both polarizations peak at a frequency corresponding to the velocity $v = v_B + v_D$, where v_B is the term attributable to the Bragg scatterers and v_D is a term encompassing the drift and orbital velocities of the underlying gravity waves. The decorrelation times of the two polarizations are short (tens of millseconds).
2. *Whitecap scattering*: The backscatter amplitudes of the two polarizations are roughly equal and are noticeably stronger than the background Bragg scatter, particularly in HH, in which Bragg scattering is weak. In a time profile, the events may be seen to last for times on the order of seconds, but are noisy in structure and decorrelate quickly (again, in milliseconds). Doppler spectra are broad and centered at a speed noticeably higher than the Bragg speed, at or around the phase speed of the larger gravity waves.
3. *Spikes*: Spikes are strong in HH, but virtually absent in VV, with a Doppler shift higher than the Bragg shift. They last for a much shorter time than the whitecap returns (on the order of 0.1 s) but remain coherent over that time.

Each of these three regimes is assigned a Gaussian line shape, with three parameters: its power (radar cross-section), center frequency, and frequency width. Assuming that the overall spectrum is a linear combination of its components, the VV spectrum is the sum of Bragg and Whitecap lineshapes, while the HH spectrum is the sum of Bragg, Whitecap, and Spike lineshapes.

The model has been validated with experimental cliff-top radar data, for which the widths and relative amplitudes of the Gaussian lineshapes were determined using a minimization algorithm.

Other researchers have similarly identified Bragg and faster-than-Bragg components, using Gaussian lineshapes for the former, and Lorentzian and/or Voigtian lineshapes for the latter [40]. The results were reported for the case of breaking waves but in the absence of wind.

Summarizing the material presented in this section, we have focused on the classical statistical approach for the characterization of sea clutter. In the next section, we consider deterministic chaos as a possible mechanism for the nonlinear dynamics of sea clutter.

4.3 IS THERE A RADAR CLUTTER ATTRACTOR?

The *Navier–Stokes dynamical equations* are basic to the understanding of the underlying principles of fluid mechanics, including ocean physics [50]. Starting with these equations, Lorenz [51] derived an unrealistically simple model for atmospheric turbulence, which is described by three coupled nonlinear differential equations. The model, bearing his name, was obtained by deleting everything from the Navier–Stokes equations that appeared to be extraneous to the simplest mathematical description of the model. The three equations, governing the evolution of the *Lorenz model*, are deceptively simple, but the presence of certain nonlinear terms in all three equations gives rise to two unusual characteristics:

(i) A *fractal dimension*[3] equal to 2.01.

(ii) *Sensitivity to initial conditions*, meaning that a very small perturbation in initialization of the model results in a significant deviation in the model's trajectory in a relatively short interval of time.

These two properties are the hallmark of *chaos* [52], a subject that has captured the interests of applied mathematicians, physicists, and (to a much lesser extent) signal-processing researchers during the past two decades.

4.3.1 Nonlinear Dynamics

Before anything else, for a process to qualify as a chaotic process, its underlying dynamics must be *nonlinear*. One test that we can use to check for the nonlinearity of an experimental time series is to employ *surrogate data* analysis [53]. The surrogate data are generated by using a stochastic linear model with the same autocorrelation function, or equivalently power spectrum, as the given time series. The exponential growth of interpoint distances between these two models is then used as the discriminating statistic to test the *null hypothesis* that the experimental time series can be described by linearly correlated noise. For this purpose, the *Mann–Whitney rank-sum statistic*, denoted by the symbol Z, is calculated. The statistic Z is Gaussian distributed with zero mean and unit variance under the null hypothesis that two observed samples of interpoint distances calculated for the experimental time series and the surrogate time series come from the same population. A value of Z less than -3.0 is considered to be a solid reason for strong rejection of the null hypothesis, that is, the experimental time series is nonlinear [54].

[3] Ordinarily, the dimension of a model is some positive integer number. In contrast, the fractal dimension is non-integer.

Appendix A summarizes three real-life sea clutter datasets, which were collected with the IPIX radar on the East Coast of Canada. Specifically, the datasets used in the case study are as follows:

- Dataset $L1$, corresponding to a lower sea state, with the ocean waves moving away from the radar; the sampling frequency of this dataset is twice that of the other two datasets H and $L2$.
- Dataset H, corresponding to a higher sea state, with the ocean waves coming toward the radar.
- Dataset $L2$, corresponding to a lower sea state, taken earlier the same day as dataset H, and at a radar range of 4 km compared to 1.2 km for dataset H; this difference in range causes H to have a considerably better signal-to-noise ratio than $L2$.

Two different types of surrogate data were generated:

- Datasets $L1_{\text{surr1}}$, $L2_{\text{surr1}}$, and H_{surr1}, which were respectively derived from sea clutter datasets $L1$, $L2$, and H using the Tisean package described in reference 55. The method keeps the linear properties of the data specified by the squared amplitude of its Fourier transform, but randomizes higher order properties by shuffling the phases.
- Datasets $L1_{\text{surr2}}$ and H_{surr2}, which are respectively derived from sea clutter datasets $L1$ and H using a procedure based on the compound K-distribution, as described in Conte et al. [43].

Table 4.1 summarizes the results of applying the Z-test to the real-life and surrogate datasets. The calculations involving surrogate datasets $L1_{\text{surr1}}$, $L2_{\text{surr1}}$, and H_{surr1} were repeated four times with different random seeds, to get a feel for the variability of the Z-statistic (and chaotic invariants to be discussed). Based on the results summarized in Table 4.1, we may state the following:

1. Sea clutter is *nonlinear* (i.e., Z is less than –3) when the underlying sea state is high.
2. No evidence for nonlinearity is found (i.e., Z is larger than –3) when the sea state is low and the ocean waves are moving away from the radar. But, when the ocean waves are coming toward the radar, sea clutter is *nonlinear* (i.e., Z is less than –3) even when the sea state is low.
3. The surrogate datasets have larger Z-values (i.e., less evidence for nonlinearity) than their original counterparts. (Surrogate datasets $L1_{\text{surr1}}$, $L2_{\text{surr1}}$ and H_{surr1} are linear by construction.)

On the basis of these results, we may state that sea clutter is a nonlinear dynamic process, with the nonlinearity depending on the sea state being moderate or higher and the ocean waves moving toward or away from the radar.

The next question to be discussed is whether sea clutter is close to being deterministic chaotic. If so, we may then apply the powerful concepts of chaos theory.

4.3.2 Chaotic Invariants

In the context of a chaotic process, two principal features, namely, the correlation dimension and Lyapunov exponents, have emerged as *invariants*, with each one of them highlighting a distinctive characteristic of the process.

Physical processes that require an external source of energy are dissipative. For sea clutter, wind and temperature differences (caused by solar radiation) are the external sources of energy. A dissipative chaotic system is characterized by its own attractor. Consider then the set of all admissible initial conditions in the multidimensional state-space of the system, and call this the initial volume. The existence of an attractor implies that the initial volume eventually collapses onto a geometric region whose dimensionality is smaller than that of the original state space. Typically, the attractor has a multisheet structure that arises from the interplay between stabilizing and disrupting forces. The *correlation dimension*, originated by Grassberger and Procaccia [12], provides an invariant measure of the geometry of the attractor. For a chaotic process, the correlation dimension is always fractal (i.e., non-integer).

Whereas the correlation dimension characterizes the distribution of points in the state space of the attractor, the *Lyapunov exponents* describe the action of the dynamics defining the evolution of the attractor's trajectories. Suppose that we now picture a small sphere of initial conditions around a point in the states pace of the attractor and then allow each initial condition to evolve in accordance with the nonlinear dynamics of the attractor; then we find that in the course of time, the small sphere of initial conditions evolves into an ellipsoid. The Lyapunov exponents measure the exponential rate of growth or shrinkage of the principal axes of the evolving ellipsoid. For a process to be chaotic, at least one of the Lyapunov exponents must be positive so as to satisfy the requirement of sensitivity to initial conditions. Moreover, the sum of all Lyapunov exponents must be negative so as to satisfy the dissipative requirement.

With this brief overview of chaotic dynamics, we return to the subject at hand: the nonlinear dynamics of sea clutter. In an article published in 1990, Leung and Haykin [19] posed the following question: "Is there a radar clutter attractor?" By applying the Grassberger–Procaccia algorithm to sea clutter, Leung and Haykin obtained a fractal dimension between 6 and 9. Independently of this work, Palmer et al. [20] obtained a value between 5 and 8 for the correlation dimension of sea clutter.

These initial findings prompted Haykin and co-investigators to probe more deeply into the possible characterization of sea clutter as a chaotic process by looking into the second invariant: Lyapunov exponents. Haykin and Li [22] reported one positive Lyapunov exponent followed by an exponent very close to zero, along with several negative exponents. This was followed by a more detailed investigation by Haykin and Puthusserypady [23], using state-of-the-art algorithms:

- A maximum-likelihood-based algorithm for estimating the correlation dimension [56].
- An algorithm based on Shannon's mutual information for measuring the embedding delay [57, 58].

- Global embedding dimension, using the method of false nearest neighbors [59]; the *embedding dimension* is defined as the smallest integer dimension that unfolds the attractor.
- Local embedding dimension, using the method of local false nearest neighbors [60]; the local embedding dimension specifies the size of the Lyapunov spectrum.
- An algorithm for estimating the Lyapunov exponents, which involves recursive QR decomposition applied to the Jacobian of a function that maps points on the trajectory of the attractor into corresponding points a prescribed number of time steps later [61, 62].

The findings reported by Haykin and Puthusserypady [23] for sea clutter are summarized here:

- Correlation dimension between 4 and 5.
- Lyapunov spectrum consisting essentially of five exponents, with two positive, one close to zero, and the remaining ones negative, with the sum of all the exponents being negative.
- Kaplan–Yorke dimension, derived from the Lyapunov spectrum, very close to the correlation dimension.

These findings were so compelling, in light of known chaos theory, that the generation of sea clutter was concluded to be the result of a chaotic mechanism, on which we have more to say in Section 4.3.5.

4.3.3 Inconclusive Experimental Results on the Chaotic Invariants of Sea Clutter

The algorithms currently available for estimating the chaotic invariants of experimental time series work very well indeed when the data are produced by mathematically derived chaotic models (e.g., the Lorenz attractor), even in the presence of additive white noise so long as the signal-to-noise ratio is moderately high. Unfortunately, they do not have the necessary discriminative power to distinguish between a deterministic chaotic process and a stochastic process. We have found this serious limitation for all the algorithms in our chaos analysis toolbox. The estimates of the correlation dimension and Lyapunov spectrum are summarized in Tables 4.1 and 4.2. Based on these results, we may make the following observations:

(i) Examining the last column of Table 4.1 on the maximum-likelihood estimate of the correlation dimension, we see that for all practical purposes there is little difference between the correlation dimension of sea clutter data and that of their surrogate counterparts that are known to be stochastic by design.

(ii) Examining Table 4.2 on the estimates of Lyapunov spectra and the derived Kaplan–Yorke dimension, we again see that a test based on Lyapunov exponents is incapable of distinguishing between the dynamics of sea clutter data and their respective stochastic surrogates.

Table 4.1 Summary of Z-Tests and Correlation Dimension

Dataset	Z-Statistic	Maximum Likelihood Estimate of Correlation Dimension
***L*1**	**0.1**	**5.1**
$L1_{\text{surr1a}}$	−0.2	5.8
$L1_{\text{surr1b}}$	0.0	5.8
$L1_{\text{surr1c}}$	−0.4	5.7
$L1_{\text{surr1d}}$	−0.4	5.6
$L1_{\text{surr2a}}$	−0.7	5.8
***L*2**	**−5.4**	**5.9**
$L2_{\text{surr1a}}$	−2.8	5.2
$L2_{\text{surr1b}}$	−3.7	5.2
$L2_{\text{surr1c}}$	−3.4	5.5
$L2_{\text{surr1d}}$	−2.9	5.4
H	**−3.5**	**4.4**
H_{surr1a}	−0.5	5.4
H_{surr1b}	0.3	5.3
H_{surr1c}	−1.1	5.3
H_{surr1d}	−0.1	5.3
H_{surr2}	−2.9	4.8

Similar results on the inadequate discriminative power of some of these algorithms are reported in reference 63.

We thus conclude that although sea clutter is nonlinear, its chaotic invariants are essentially the same as those of the surrogates that are known to be stochastic. The notion of nonlinearity alone does not imply deterministic chaos; it merely excludes the possibility that a linear mechanism is responsible for the generation of sea clutter.

Note that dataset *L*1 was sampled at 2 kHz, while datasets *L*2 and *H* were sampled at 1 kHz. Therefore, the *L*1 horizon values listed in the table appear as approximately double the values for *L*2.

4.3.4 Dynamic Reconstruction

All along, the driving force for the work done by Haykin and co-investigators has been the formulation of a *robust dynamic reconstruction algorithm* to make physical

Table 4.2 Summary of Lyapunov Exponents[a]

Set	Exp 1	Exp 2	Exp 3	Exp 4	Exp 5	Exp sum	Dimen	Horiz
***L*1**	**0.1046**	**0.0411**	**−0.0154**	**−0.0864**	**−0.2670**	**−0.2231**	**4.16**	**37.41**
$L1_{\text{surr1a}}$	0.1184	0.0419	−0.0196	−0.1028	−0.3003	−0.2625	4.13	33.03
$L1_{\text{surr1b}}$	0.1176	0.0443	−0.0129	−0.1003	−0.3130	−0.2643	4.16	33.26
$L1_{\text{surr1c}}$	0.1152	0.0457	−0.0198	−0.1045	−0.2876	−0.2509	4.13	33.95
$L1_{\text{surr1d}}$	0.1222	0.0473	−0.0149	−0.0947	−0.2867	−0.2269	4.21	32.02
$L1_{\text{surr2}}$	0.1211	0.0445	−0.0235	−0.1163	−0.3418	−0.3159	4.08	32.30
***L*2**	**0.4472**	**0.2675**	**0.0606**	**−0.2448**	**−0.8735**	**0.3429**	**4.61**	**8.75**
$L2_{\text{surr1a}}$	0.2304	0.1032	0.0315	0.2185	−0.6447	0.5611	4.13	16.98
$L2_{\text{surr1b}}$	0.2395	0.1143	0.0235	0.2020	−0.6762	0.5479	4.19	16.34
$L2_{\text{surr1c}}$	0.2352	0.1088	0.0267−	0.1970−	0.6847	0.5645	4.18	16.63
$L2_{\text{surr1d}}$	0.2434	0.1176	0.0245−	0.1876	0.6708	0.5218	4.22	16.07
H	**0.4058**	**0.2405**	**0.0644**	**0.1998**	**0.7674**	**0.2565**	**4.67**	**9.64**
H_{surr1a}	0.3521	0.1871	0.0193	0.2107	0.8399	0.4921	4.41	11.11
H_{surr1b}	0.3836	0.2066	0.0221	0.2092	0.7885	0.3853	4.51	10.20
H_{surr1c}	0.3877	0.2038	0.0160	0.2111	0.7994	0.4028	4.50	10.09
H_{surr1d}	0.3670	0.1941	0.0085	0.2008	0.8027	0.4339	4.46	10.66
H_{surr2}	0.3556	0.2007	0.0240	0.2248	0.7917	0.4361	4.45	11.00

[a] Notations used in Table 4.2: Exp 1 to Exp 5 denote the estimated Lyapunov exponents, given in units of nats per sample. Exp sum denotes the sum of the Lyapunov exponents. Dimen denotes the Kaplan–Yorke dimension defined by

$$D_{KY} = k + \frac{\lambda_1 + \ldots + \lambda_k}{|\lambda_{k+1}|}, \qquad \lambda_1 > \lambda_2 > \ldots > \lambda_k > \ldots, k = \max\{i, \lambda_1 + \ldots + \lambda_i > 0\}$$

Horiz denotes the horizon of predictability, which is computed from the Lyapunov exponents. The horizon of predictability is given in units of sample period, which can be converted to time in seconds by dividing by the sampling rate.

sense of real-life sea clutter by capturing its underlying dynamics. Such an algorithm is essential for the reliable modeling of sea clutter and the improved detection of a target in sea clutter. Successful development of such a dynamic reconstruction algorithm was also considered to be further evidence of deterministic chaos as the descriptor of sea clutter dynamics.

To describe the dynamic reconstruction problem, with chaos theory in mind, consider an attractor whose process equation is noiseless and whose measurement noise is additive, as shown by the following pair of equations:

$$\mathbf{x}_{t+1} = \mathbf{f}(\mathbf{x}_t) \tag{4.9}$$

$$y_t = h(\mathbf{x}_t) + w_t \tag{4.10}$$

Suppose that we use the set of noisy observations $\{y_t\}_{t=1}^{N}$ to construct the vector

$$\mathbf{r}_t = [y_t, y_{t-\tau}, \ldots, y_{t-(D-1)\tau}]^T \tag{4.11}$$

where τ is the *embedding delay* equal to an integer number of time units, D is the embedding dimension, and the superscript T denotes matrix transposition. As the observations evolve in time, the vector $\mathbf{r}_t$ defines the underlying attractor, thereby providing a fiducial trajectory. The stage is now set for stating the *delay-embedding theorem* due to Takens [10], Mañé [11], and Sauer et al. [16]:

> *Given the experimental time series $\{y_t\}$ of a single scalar component of a nonlinear, finite-dimensional dynamic system, the geometric structure of the hidden dynamics of that system can be unfolded in a topologically equivalent manner in that the evolution of the points $\mathbf{r}_t \to \mathbf{r}_{t+1}$ in the reconstructed state space follows the evolution of the points $\mathbf{x}_t \to \mathbf{x}_{t+1}$ in the original state space, provided that D is equal or greater than $2D_0 + 1$, where D_0 is the fractal dimension of the system and the vector $\mathbf{r}_t$ is related to the given time series $\{y_t\}$ by (4.11).*

Ideas leading to the formulation of the delay-embedding theorem were described in an earlier paper by Packard et al. [9].

A key point to note here is that since all the variables of the system are geometrically related to each other in a nonlinear manner, as shown in (4.9) and (4.10), measurements made on a single component of the nonlinear dynamic system contain sufficient information to reconstruct the multidimensional state $\mathbf{x}_t$.

Derivation of the delay-embedding theorem rests on two key assumptions:

- The model is noiseless; that is, not only is the state equation (4.9) noiseless, but the measurement equation (4.10) is also noiseless (i.e., $w_t = 0$).
- The observable dataset $\{y_t\}$ is infinitely long.

Under these conditions, the theorem works with any delay τ so long as the embedding dimension D is large enough to unfold the underlying dynamics of the process of interest.

Nevertheless, given the reality of a noisy dynamic model described by (4.9) and (4.10) and given a finite record of observations $\{y_t\}_{t=1}^{N}$, the delay-embedding theorem may be applied, provided that a "reliable" method is used for estimating the embedding delay τ. According to Abarbanel [64], the recommended method is to compute that particular τ for which the mutual information between $\{y_t\}$ and its delayed version $\{y_{t-\tau}\}$ attains its minimum value; and the recommended method for estimating the embedding dimensions D_0 is to use the method of false nearest neighbors.

A distinction must be made between dynamic reconstruction and predictive modeling. *Predictive modeling* is an open-loop operation, which merely requires that the prediction error (i.e., the difference between the present value of a time series and its nonlinear prediction based on a prescribed set of past values of the time series) be minimized in the mean-square sense. Dynamic reconstruction is more demanding, in that its builds on a predictive model by requiring closed-loop operation. Specifically, the predictive model is initialized with data drawn from the same process under study but not seen before, and then the model's output is delayed by one time unit and fed back to the input layer of the model, making room for this new input sample by leaving out the oldest sample in the initializing data set. This procedure is continued until the entire initializing dataset is completely disposed of. Thereafter, the model operates in an *autonomous* manner, producing an output time series learned from the data during the training (i.e., open-loop predictive) session.

It is amazing that dynamic reconstruction, as described herein, works well for time series derived from mathematical models of deterministic chaos, even when the time series is purposely contaminated with additive white noise of relatively moderate average power (see, for example, reference 65).

Unfortunately, despite the persistent use of different reconstruction procedures involving the use of a multilayer perceptron trained with the back-propagation algorithm [22], regularized radial-basis function (RBF) networks [66], and recurrent multilayer perceptrons trained with the extended Kalman filter [65], the formulation of a reliable procedure for the dynamic reconstruction of sea clutter based on the delay-embedding theorem that would work all of the time, not just some of the time, has eluded us. The key question is, Why? A feasible answer to this important question is offered in Section 4.3.5.

Serious difficulties with the dynamic reconstruction of sea clutter prompted the authors of this chapter in September 2000 to question the validity of a chaotic model for describing the nonlinear dynamics of sea clutter, despite the highly encouraging results summarized in Section 4.3. Indeed, it was because of these serious concerns that a complete reexamination of the nonlinear dynamic modeling of sea clutter was undertaken, as detailed in Section 4.4. However, before moving onto that section, we conclude the present discussion on chaos by highlighting some important *lessons* learned from our work on the application of deterministic chaos to sea clutter.

4.3.5 Chaos, a Self-Fulfilling Prophecy?

Chaos theory provides the mathematical basis of an elegant discipline for explaining complex physical phenomena using relatively simple nonlinear dynamic models. As with every scientific discipline that requires experimentation with real-life data, we clearly need *reliable* algorithms for estimating the basic parameters that characterize the physical phenomenon of interest, given an experimental time series. As already mentioned, there are two invariants that are basic to the characterization of a chaotic process:

- Correlation dimension
- Lyapunov exponents

Unfortunately, state-of-the-art algorithms for estimating these invariants do not have the necessary *discriminative power* to distinguish between a deterministic chaotic process and a stochastic process. For the experimenter who hopes his/her data qualify for a deterministic chaotic model, the results of a chaotic invariant analysis may end up working as a "self-fulfilling prophecy," indicating the existence of deterministic chaos regardless of whether the data are really chaotic or not. The stochastic process could be colored noise, or a nonlinear dynamic process whose state-space model includes dynamic noise in the process equation. As pointed out in Sugihara [67], when we have noise in both the process and measurement equations of a nonlinear dynamic model, there is unavoidable practical difficulty in disentangling the dynamic (process) noise from the measurement noise to reconstruct an invariant measure. Specifically, in the estimation of Lyapunov exponents, it is no longer possible to compute meaningful products of Jacobians from the experimental time series because the invariant measure is contaminated with noise.

We now turn to the fundamental question at hand: How do we explain the possible presence of dynamic noise in the state-space model of sea clutter? To answer this important question, we first need to remind ourselves that ocean dynamics are affected by a variety of forces, as summarized here [50]:

- Gravitational and rotational forces, which permeate the entire fluid, with large scales compared with most other forces.
- Thermodynamic forces, such as radiative transfer, heating, cooling, precipitation, and evaporation.
- Mechanical forces, such as surface wind stress, atmospheric pressure variations, and other mechanical perturbations.
- Internal forces—pressure and viscosity—exerted by one portion of the fluid on other parts.

With all these forces acting on the ocean dynamics, and therefore directly or indirectly influencing radar backscatter from the ocean surface, three effects arise:

1. Evolution of the hidden state characterizing the underlying dynamics of sea clutter due to the constant state of motion of the ocean surface.
2. Generation of some form of dynamic noise, contaminating evolution of the state with time, due to the natural rate of variability of the forces acting on the ocean surface.
3. Imposition of a nonstationary spatio-temporal structure on the radar observable(s).

Hence, we have to face the physical reality that in addition to measurement noise, there is dynamic noise to deal with. Moreover, the fact that there is usually no prior knowledge of the measurement noise or dynamic noise, it is not surprising that the dynamic reconstruction of sea clutter using experimental time series is a very difficult proposition indeed.

4.4 HYBRID AM/FM MODEL OF SEA CLUTTER

In Section 4.3, we have expressed doubts on the validity of deterministic chaos approach as a descriptor of sea clutter. In this section, we take a detailed look at several experimental datasets to explore new ways to model sea clutter. Apart from the classical approaches reviewed in Section 4.2, one source of inspiration is the recent work of Gini and Greco [68], who view sea clutter as a fast "speckle" process, multiplied by a "texture" component that represents the slowly varying mean-power level of the data—caused by large waves passing though the observed ocean patch; they model the speckle as a stationary compound complex Gaussian process, and they model the texture as a harmonic process. What we will find is that the relationship between the slow- and fast-varying processes is much more involved than what has been assumed so far in the literature. In particular, we find that the slowly varying component modulates not only the amplitude of the speckle, but also its mean frequency and spectral width.

4.4.1 Radar Return Plots

We use the datasets *L*2 and *H* from Appendix A. The analysis starts by looking at the radar return plots of these datasets, shown in Fig. 4.5. These plots show the strength of the radar return signal (color axis) as a function of time (x-axis) and range (y-axis). A dark color indicates a strong return, which is associated with wave crests. The diagonal dark stripes in both plots show that the wave crests move, with increasing time, toward a decreasing range or toward the radar. Indeed, we see from Appendix A that the wind and radar beam point in almost opposite directions. If we look at a single range bin (i.e., along a horizontal line in the radar return plots), we see that the strength of the return signal is roughly periodic, with a period in the order of 4 to 8 s, corresponding to the period of the gravity waves (see Section 4.2). In Figs. 4.1a–4.1c, such a single range bin is plotted against time, the y-axis now being return strength (amplitude of the received signal). The periodic behavior is less pronounced due to the wild short-term fluctuations of the signal, which are caused by Rayleigh fading.

4.4.2 Rayleigh Fading

Rayleigh fading arises when a number of complex exponentials of slightly different frequencies are added together. Figure 4.6 shows the magnitude and instantaneous frequency of the sum $x = a_1 \exp(j2\pi f_1 t) + a_2 \exp(i2\pi f_2 t)$, where $f_1 = 1$, $f_2 = 1.1$, $a_1 = 1$, and a_2 is varied; j denotes the square root of -1. In Fig. 4.6a, a_2 is equal to a_1; in Fig. 4.6b, it is 10% larger. The figure illustrates the typical upside down U shape of the magnitude of a Rayleigh fading process, with a period T_{Rayleigh} that follows.

$$T_{\text{Rayleigh}} = 1/|f_1 - f_2| \tag{4.12}$$

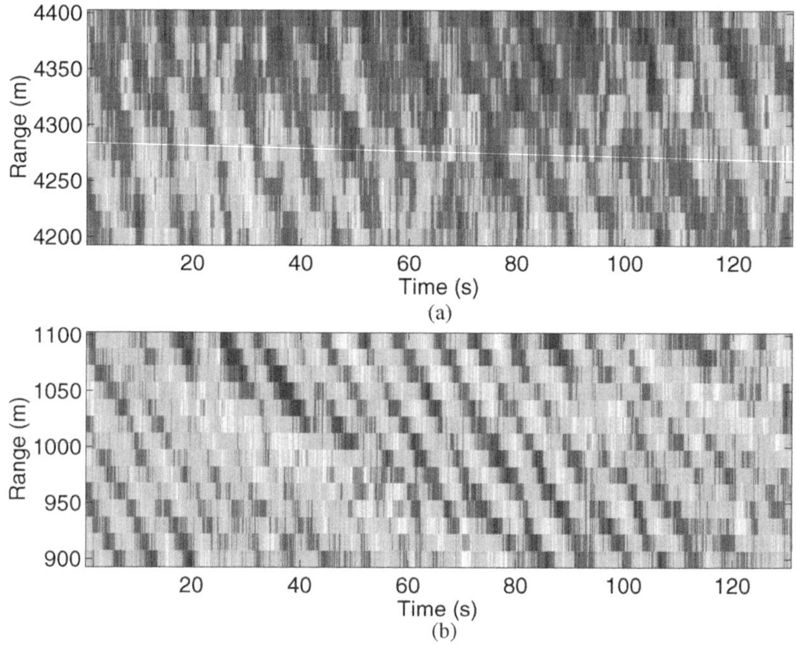

Figure 4.5 Radar return versus time and range, VV polarization, of (a) dataset *L2* (low sea state) and (b) dataset *H* (high sea state). The color axis shows $\log(|\tilde{x}|)$, where $\tilde{x}$ is the complex envelope of the received signal. The units of $\tilde{x}$ are normalized and the color axis changes from blue (low) via green to red (high). See color insert.

Looking at the close-up of our data in Fig. 4.7, we see that both the magnitude and instantaneous frequency have the typical characteristics of Rayleigh fading, although in Section 4.4.3 we will find that most spikes in the deviation $\dot{\phi} = d\phi/dt$ time series are actually caused by receiver noise.

Why does Rayleigh fading occur? The answer lies in the independent scatterer model that Jakeman and Pusey [27] first used to derive a physical justification for the use of the *K*-distribution (see Section 4.2). If we think of a patch of ocean illuminated by the radar at a given time, according to this model the received signal will be dominated by a small number of independent scatterers, each moving at its own velocity. We make the additional restriction that, at least for the short duration of a single sample time, each scatterer has its own constant velocity with respect to the radar. The received signal may then be modeled as

$$y(t) = \exp(j\omega_{\mathrm{RF}}t)\sum_{k=1}^{N} a_k \exp(j\omega_{D,k}(t-t_0)+\phi_{t_0,k}) \tag{4.13}$$

where ω_{RF} is the RF angular frequency of the radar (equal to 2π times 9.39 GHz for the IPIX radar), N is the number of independent scatterers, a_k is proportional to the

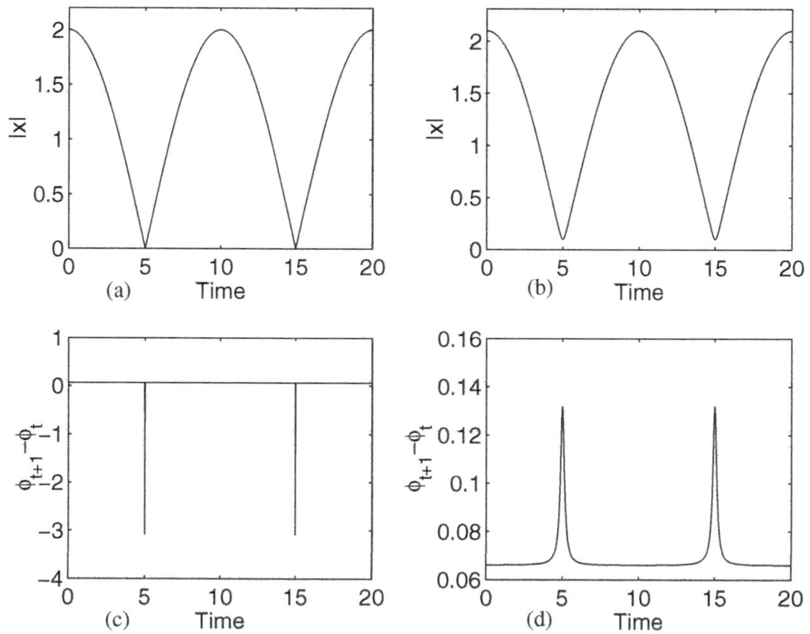

Figure 4.6 Magnitude (a, b) and instant aneous frequency (c, d) of the sum of two complex exponentials: $\exp(j2\pi t) + \exp(j2\pi 1.1t)$ (a, c); $\exp(j2\pi t) + 1.1\ \exp(j2\pi 1.1t)$ (b, d). Instantaneous frequency is computed as the difference between the phase of subsequent samples, after unwrapping the phase to remove 2π jumps.

effective radar cross-section of scatterer k, and $\omega_{D,k}$ and $\phi_{t_0,k}$ are respectively the angular Doppler frequency and phase of scatterer k at time t_0. After removing the carrier wave by multiplying by $e^{-j\omega_{RF}t}$, we see that (4.13) is indeed a sum of complex exponentials with slightly different frequencies, thereby resulting in Rayleigh fading. The frequencies are related to the physical speed of the scatterer by (4.8).

Now that we have established the Rayleigh fading characteristics of sea clutter amplitude data, we go back to (4.12), which relates, for the case of two complex exponentials, the period of the amplitude signal to the frequency difference of the two exponentials. Sea clutter has much more than two complex exponentials, and they are of constantly changing frequency. But as a coarse approximation, (4.12) may still be useful. As a rough estimate for the average T_{Rayleigh} of sea clutter, we use the *average cycle time* (ACT) of the data: Subtract the median of the signal and then take the average time between two upward zero-crossings. For the left-hand term of (4.12), we need to estimate $|f_1 - f_2|$, the frequency variability of sea clutter. This we estimate by taking the measured instantaneous frequency and computing its *normalized median absolute deviation* (NMAD). The NMAD is a robust estimate of the signal's standard deviation, ignoring the spikes. It is computed as

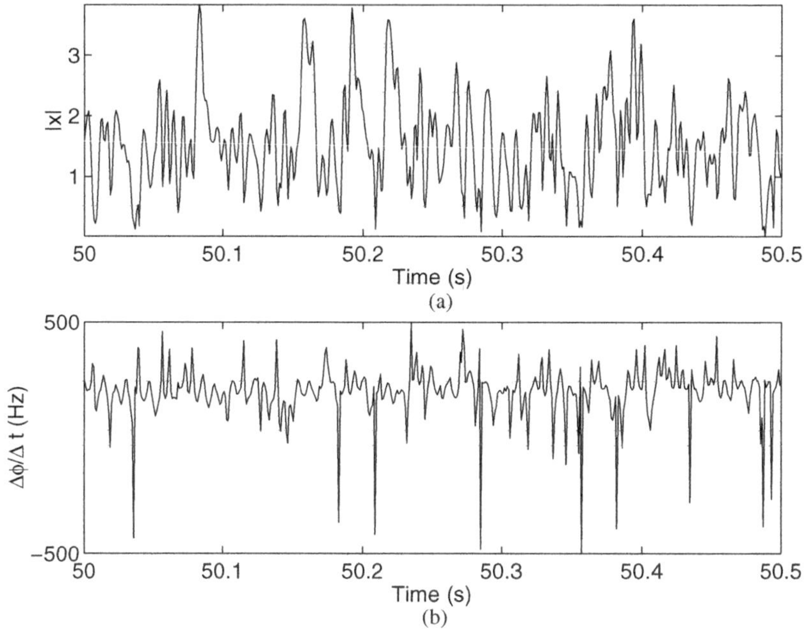

Figure 4.7 (a) Close-up of the radar return signal of Fig. 4.11a. (b) Corresponding instantaneous frequency $\dot{\phi}_t$, as related to Fig. 4.11b. It is computed $(\phi_{t+1} - \phi_t)/(2\pi\Delta t)$, where the phase ϕ is first unwrapped to remove jumps larger than π, and Δt is the sampling time of 1 ms.

$$\text{NMAD}(\dot{\phi}) = 1.48 \times \text{median}\left(\left|\dot{\phi} - \text{median}\left(\dot{\phi}\right)\right|\right) \tag{4.14}$$

For the example of Fig. 4.7, we have $\text{ACT}(|x|) = 0.01$ s, and $\text{NMAD}(\dot{\phi}) = 54$ Hz. If we take twice the standard deviation as our measure of variability, then the result satisfies (4.12), as $0.01 \approx 1/(2 \times 54)$. In Fig. 4.8, we look at how the two quantities $2\text{NMAD}(\dot{\phi})$ and $1/\text{ACT}(|x|)$ evolve with time. We use a moving window of 1000 samples (1 s). For the high sea state, the two curves almost overlap, in agreement with (4.12). For the low sea state, the curves do not overlap but they follow the same trends.

It is interesting to also try to link the variability of $\dot{\phi}$ to $\dot{\phi}$ itself. If we could do this, then even with an inexpensive noncoherent radar, using only the envelope of the received signal, the radar could provide a rough estimate of the speed of the observed waves. This is not the focus of this chapter, but the strong correlation seen in Fig. 4.11d shows the possible viability of this approach; the pursuit of such an application of noncoherent radar merits further investigation.

4.4.3 Time-Doppler Spectra

Since the independent scatterer model tells us that the received signal is the sum of a number of complex exponentials, it is most appropriate to describe the signal

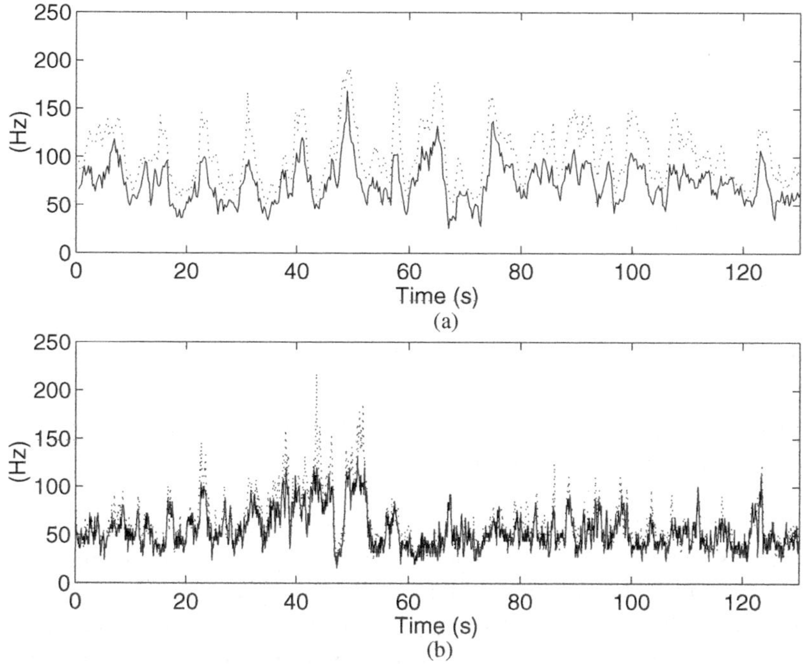

Figure 4.8 $1/\text{ACT}(|\tilde{x}|)$ (solid line) and $2\text{NMAD}(\dot{\phi})$ (dotted line) versus time, computed on a 1000-sample sliding window basis, for (a) dataset *L*2 and (b) dataset *H*.

in terms of its Fourier spectrum. But, as waves move along the observed ocean patch, we expect the number and strength of the scatterers to vary. Therefore, we again use a sliding window, this time of length 512 (0.5 s) to compute a time-varying frequency spectrum. When the frequency is converted into Doppler velocity using (4.8), it becomes the time-Doppler spectra of Figs. 4.2a–4.2c. The plots are very revealing, showing that the Doppler frequency fluctuations are a lot stronger for the higher sea-state clutter. Note also how the spectral width in the time-Doppler plots varies with time; this variation follows the same trend as the NMAD$\dot{\phi}$ signal of (4.14), which was introduced in Section 4.4.2 on Rayleigh fading.

The spectrogram contains many frequencies that are only activated by the receiver (i.e., measurement) noise part of the data. We estimated the receiver noise level by comparing the total power of the signal to the power in the part of the Doppler spectrum below −4 m/s. The signal-to-noise ratio was found to be 17 dB for dataset *L*2 and 31 dB for dataset *H*, the difference being caused by the difference in range and the reduced overall power of lower sea-state clutter.

The noise estimates are very useful to estimate the variance of the signals we derived from the data. For example, we can see immediately that most of the spikes in the $\dot{\phi}$ time series occur when the signal drops below the noise floor. And what is the variance of the $\dot{\phi}$ signal? If the magnitude of the signal is well above the noise floor, then the standard deviation $\sigma_{\phi_{t+1}-\phi_t}$ can be estimated using the formula (see Fig. 4.9)

$$\sigma_{\phi_{t+1}-\phi_t} = \sqrt{\frac{1}{|x_{t+1}|^2} + \frac{1}{|x_t|^2}}\sigma_\theta \tag{4.15}$$

where σ_θ is the standard deviation of receiver noise. From this estimate, it appears that the NMAD($\dot{\phi}$) signal in Fig. 4.8a (dataset *L*2) is dominated by receiver noise, whereas in Fig. 4.8b (dataset *H*) it is dominated by the clutter signal. For dataset *L*2, this means that we may postulate an inverse relationship between NMAD($\dot{\phi}$) and the amplitude of the signal. This relationship is clearly visible if we compare the solid line in Fig. 4.11a to the dotted line in Fig. 4.11b.

4.4.4 Evidence for Amplitude Modulation, Frequency Modulation, and More

Almost invariably, models for sea clutter distinguish between the slow time scale of the gravity waves and the fast time cale of the capillary waves. A typical approach is that of Conte et al. [43], consisting of a colored noise process that is amplitude-modulated by a slowly varying intensity component. Figure 4.10 shows the time-Doppler spectrum for such data (courtesy Alan Thomson). The results in the previous sections teach us that there is a much more intricate relationship between the fast- and slow-varying processes involved in the generation of sea clutter.

When a large wave passes through the ocean patch under surveillance, it will first accelerate and then decelerate the water on the ocean surface. The tilting of the ocean surface by the wave causes amplitude modulation. Even if scatterers arise

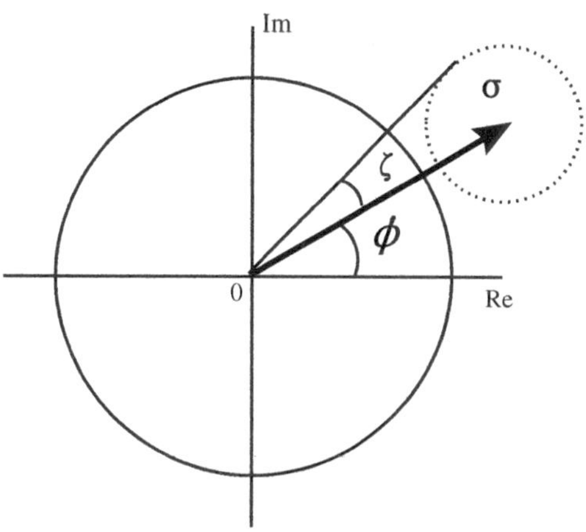

Figure 4.9 If the magnitude of the amplitude signal is high compared to the receiver noise level, the error ζ in the estimation of the angle ϕ can be approximated by the tangential component of the receiver noise, divided by the magnitude of the signal. If the receiver noise is uncorrelated, the variance of $\phi_{t+1} - \phi_t$ is $2\sigma_\phi^2$.

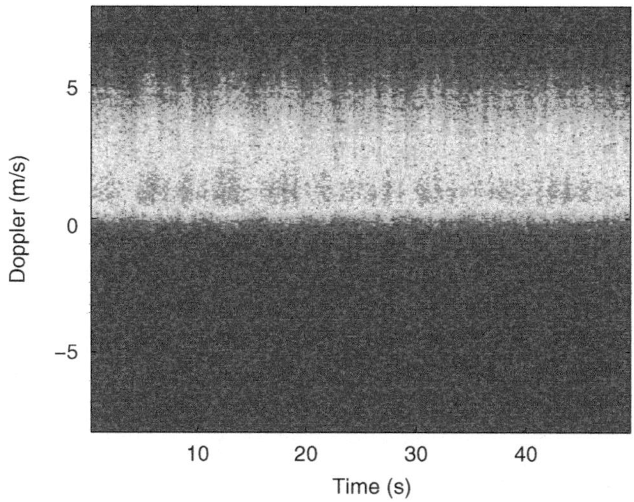

Figure 4.10 Time-Doppler spectrum of 50 s of data synthesized with the method of Conte, Longo and Lops [43]. (Data provided by Alan Thomson, DRDC, Ottawa.) See color insert.

mostly on the crest of the wave, the wave will cause a cyclic motion of the velocity of the scatterers. This motion is widely recognized, but its consequence, namely, a frequency modulation of the speckle component, has been neglected. And there is indeed more to think about.

When the mean velocity of the scatterers is high at a given instant, then the spread around that mean is also high. We can now explain why almost invariably we found sea clutter to be nonlinear when we presented the results in Table 4.1. Recognizing that amplitude modulation (in a loose sense) is a linear form of modulation but frequency modulation is not [69], we expect the value of parameter Z (discussed in Section 4.3.1) to become more negative as the amount of frequency modulation increases. Figure 4.12 confirms this qualitative relationship, by plotting the Z-value versus the amount of frequency modulation. Over and above the datasets $L1$, $L2$, and H, the figure uses an additional 75 datasets from our sea-clutter database, measured at a wide variety of experimental conditions. It is no surprise that the surrogate datasets used in Table 4.1 are less nonlinear than their original counterparts, since the random phase-shifting partially destroys the frequency modulation. The typical "breathing" phenomenon seen in time-Doppler plots, such as the ones in Fig. 4.2, shows that not only the mean of the velocity spectrum, but also its spectral width, is modulated. Moreover, in some cases, the velocity spectrum even has a bimodal distribution (around time = 45–50 s in Fig. 4.2b); the recent work of Walker [48] shows that this is most likely caused by a breaking wave.

So far we have identified four different processes acting on the dynamics of the speckle component: amplitude modulation, frequency modulation, spectral-width modulation, and bimodal frequency distributions due to breaking waves. All these four processes, which have the slow time-scale of gravity waves, need to be specified in order to synthesize artificial radar data. In Fig. 4.11, we look for

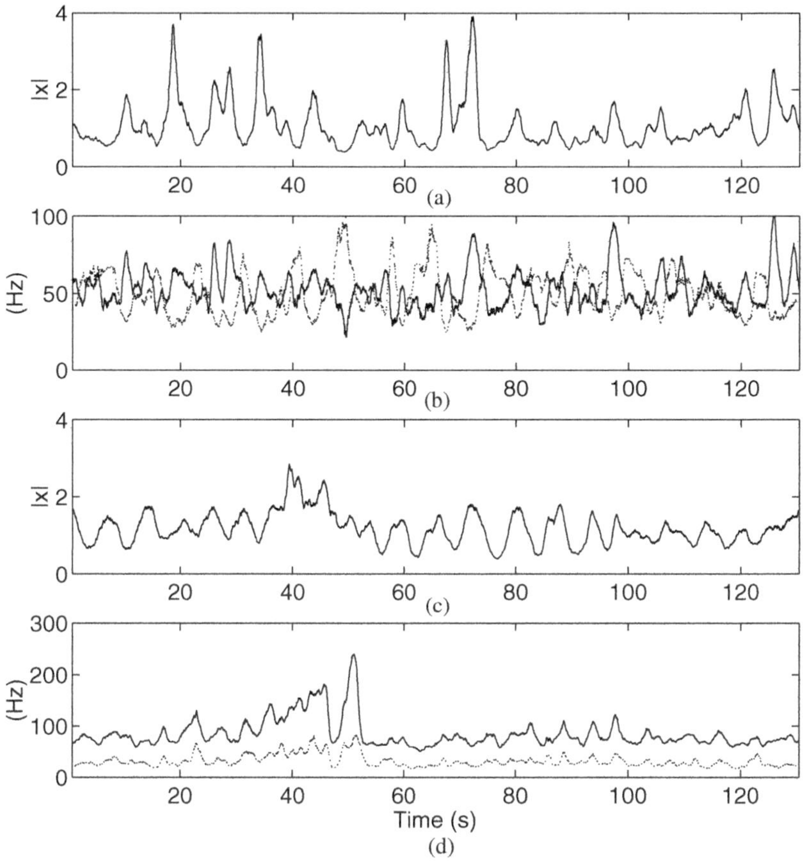

Figure 4.11 (a, c) Low-pass filtered amplitude (1-s averaging). (b, d) Low-pass filtered instant frequency $\dot{\phi}$ (solid line) and NMAD($\dot{\phi}$) (dotted line). (a, b) dataset *L*2. (c, d) dataset *H*.

possible correlation between the various types of modulation. The relation between the amplitude modulation and frequency modulation seems weak (compare solid lines in Fig. 4.11a to those in Fig. 4.11b, and compare solid lines in Fig. 4.11c to those in Fig. 4.11a). Figure 4.11d shows that there is a strong correlation between the frequency modulation ($\dot{\phi}$ averaged over 1 s) and the spectral-width modulation, measured by NMAD($\dot{\phi}$). This correlation is not confirmed by the equivalent plots for low sea state in Fig. 4.11b; but as argued in Section 4.4.3, this may be due to receiver noise.

4.4.5 Modeling Sea Clutter as a Nonstationary Complex Autoregressive Process

So far our results have not made the sea-clutter synthesis much easier—it seems we almost have to provide the entire time-Doppler spectrum to get a complete signature

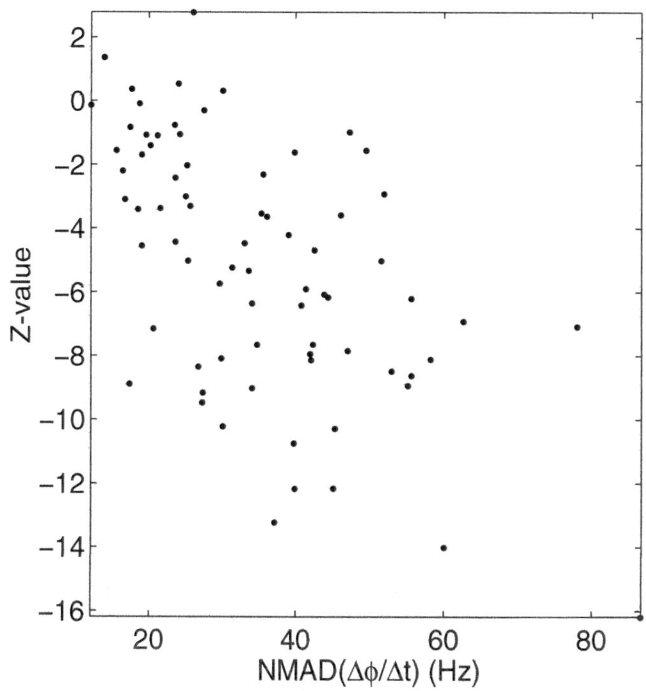

Figure 4.12 Z-value versus NMAD(ϕ), computed for 78 datasets measured by the IPIX radar at various experimental conditions.

of the observed data. As a first step toward a practical algorithm, in this section we compress the time-Doppler spectrum into only a few complex parameters per time slot, and at the same time we make it suitable for time-series generation. We argue as follows, using observations from the preceding discussion:

1. On time-scales shorter than several seconds, sea clutter can be described as the sum of complex exponentials.
2. A sum of complex exponentials is well described in terms of its Fourier spectrum.
3. The Fourier spectrum of a dynamic system can be approximated by the spectrum of an *autoregressive (AR) process*: The higher the order of the AR process, the more accurate the approximation will be. (In a sense, this is a statement of *Wold's decomposition theorem* in statistical signal processing.)

This brings us to the concept of a time-varying complex AR process. We take a 1-s window (1000 samples), slide it through dataset H (pertaining to a higher sea state) with small time increments, and each time we fit a complex AR process to the data. We search for the lowest-order time-varying AR model that approximates the short-time Fourier transforms (vertical lines in the time-Doppler spectrum) well. When we increase the order from one to four, the standard deviation of the residual error,

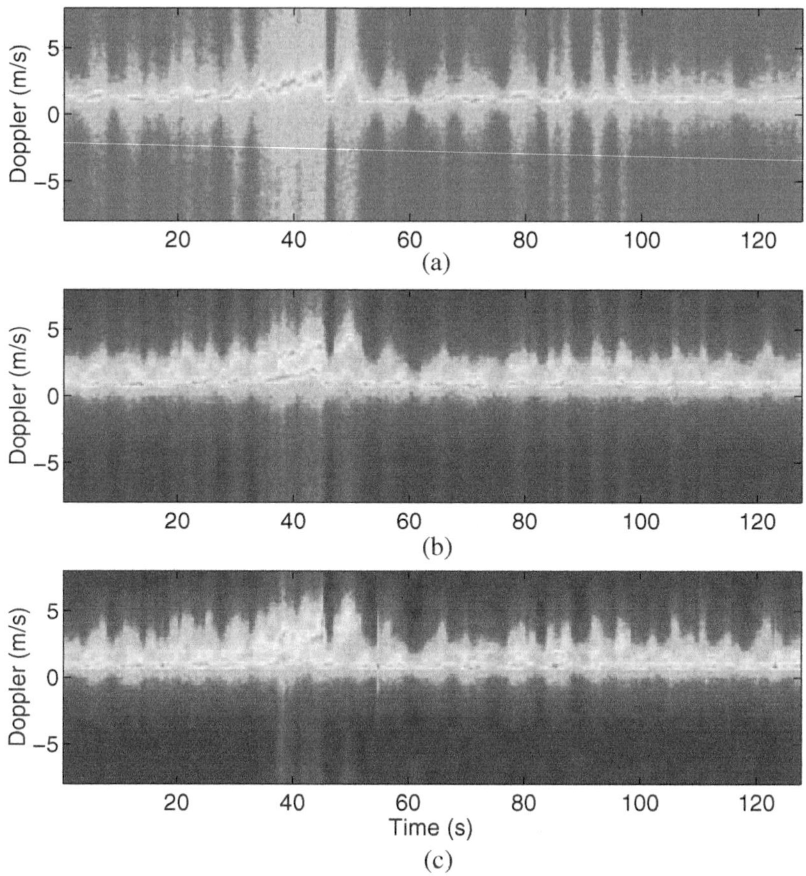

Figure 4.13 Time-Doppler spectra of data synthesized by a sliding AR process of order 1 (a), order 2 (b), and order 3 (c). The three plots of the figure have identical color axis limits. The lighter background color in plot (a) is caused by the larger residual error of the sliding AR(1) model. See color insert.

averaged over time, decreases: 0.23, 0.11, 0.091, 0.086, in units of signal standard deviations. In the same units, the receiver noise as estimated from the time-Doppler spectrum is 0.061. The improvement with model order is very clear from Fig. 4.13, which shows time-Doppler spectra of synthesized clutter, using time-varying AR processes of order 1, 2, and 3, denoted as AR(1), AR(2), and AR(3), respectively. The data are generated according to the difference equation:

$$x_{t+1} = a_{1,<t>}x_t + a_{2,<t>}x_{t-1} + \cdots + a_{K,<t>}x_{t-K+1} + e_{t_{<t>}} \tag{4.16}$$

where all variables are complex, $a_{i,<t>}$ are the AR coefficients at time $<t>$ with $i =$ 1, 2, . . . , K (the brackets indicate that they change on a slow time-scale only), K is the model order, and $e_{t,<t>}$ is the additive noise component having a time-varying variance $\sigma^2_{e_t<t>}$.

The AR model of order 1 is clearly insufficient to describe sea clutter in good detail, but it is by far the easiest to analyze in physical terms. It has three independent

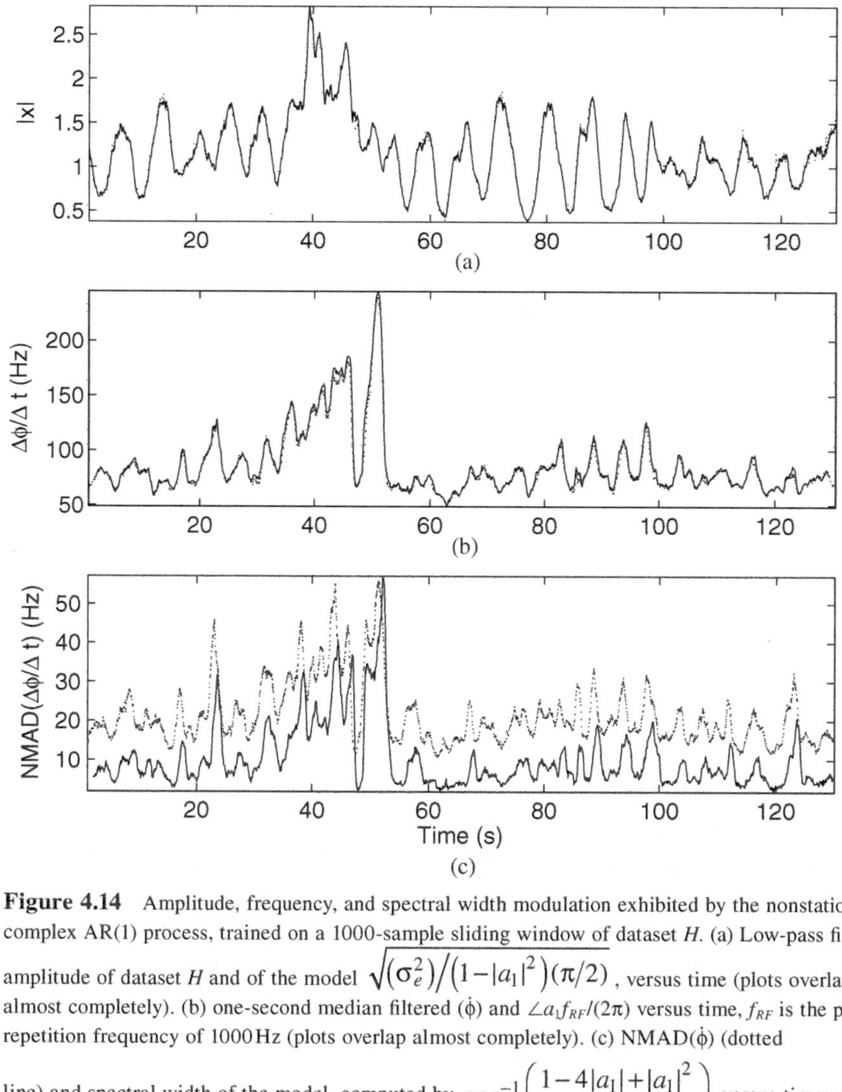

Figure 4.14 Amplitude, frequency, and spectral width modulation exhibited by the nonstationary complex AR(1) process, trained on a 1000-sample sliding window of dataset H. (a) Low-pass filtered amplitude of dataset H and of the model $\sqrt{(\sigma_e^2)/(1-|a_1|^2)(\pi/2)}$, versus time (plots overlap almost completely). (b) one-second median filtered $(\dot{\phi})$ and $\angle a_1 f_{RF}/(2\pi)$ versus time, f_{RF} is the pulse repetition frequency of 1000 Hz (plots overlap almost completely). (c) NMAD$(\dot{\phi})$ (dotted line) and spectral width of the model, computed by $\cos^{-1}\left(\dfrac{1-4|a_1|+|a_1|^2}{-2|a_1|}\right)$ versus time; see reference 70.

parameters that vary slowly with time: (1) the amplitude of $a_{1,<t>}$, (2) the angle of $a_{1,<t>}$, and (3) the variance of the noise, $\sigma^2_{e_t,<t>}$. Figure 4.14 shows how these three independent parameters are coupled to the three main types of modulation mentioned in Section 4.4.4.

Indeed, we could rewrite the nonstationary AR(1) process into an equivalent stationary AR(1) process, modulated in amplitude, frequency, and spectral width. Unfortunately, the sliding AR(1) does not provide a good enough description of the data.

4.5 DISCUSSION

Sea clutter, referring to radar backscatter from an ocean surface, is a nonstationary, complex, nonlinear dynamic process with a discernible structure that exhibits a multitude of continuous-wave modulation processes: amplitude modulation, frequency modulation, spectral-width modulation, and bimodal frequency distribution due to breaking waves. The modulations are slowly varying (on the order of seconds) functions of time. The amplitude modulation is clearly discernible in the sea-clutter waveform, regardless of the sea state or whether the ocean waves are moving away from the radar or coming toward it. The frequency modulation and variations in spectral width and spectral shape become clearly observable only when the nonlinear nature of sea clutter becomes pronounced, which happens when the sea state is high enough or the ocean waves are coming toward the radar.

4.5.1 Nonlinear Dynamics of Sea Clutter

In this chapter, we have justified the nonlinear nature of sea clutter on two grounds:

- Experimentally, applying the *Z*-test to real-life radar data, as demonstrated in Table 4.1.
- Physically, by demonstrating the existence of frequency modulation (a nonlinear process) in the composition of sea clutter, as explained in subsection 4.4.4.

Figure 4.12 links the degree of nonlinearity of sea clutter to the extent of frequency modulation in a rather vivid manner. Specifically, the higher the sea state, the more nonlinear sea clutter would be, which is intuitively satisfying.

The issue of the dynamics of sea clutter has also been addressed in the literature independently from an entirely theoretical point of view. Perhaps most notable in this latter case, Field and Tough [71, 72] develop a theoretical basis for the dynamics of sea clutter. More precisely, given the scattering dynamics described in terms of stochastic differential equations, which are derived essentially from first principles based on Ito calculus, as described in references 71 and 72, it has been shown that a certain corresponding "noise-free skeleton" of sea clutter is a nonlinear deterministic process. The degree of nonlinearity is dependent on the sea state or "shape parameter" in a manner entirely consistent with that shown experimentally in this chapter. The detail of this significant development, which affirms the case for the nonlinear nature of sea clutter, is reported separately in reference 73.

4.5.2 Autoregressive Modeling of Sea Clutter

When the issue of interest is that of addressing a computational procedure for studying the underlying dynamics of sea clutter in a phenomenological way, the use of a time-varying complex-valued autoregressive (AR) model is attractive for the following reason:

- An AR model of relatively low order is capable of capturing the major dynamical features of sea clutter, especially so when the time scale is smaller than a few seconds. More specifically, in Chapter 5 it is shown that an AR model of order 3 offers a good compromise between model complexity and model accuracy in capturing the long-term nonstationarity of sea clutter.

Note however that we are speaking of a predictive model whose parameters are functions of time t for all t. In reality, such a model is nonlinear as it violates the principle of superposition.

It is also noteworthy that, starting in the mid-1970s and for a good part of the 1980s, the first author of this chapter and other co-investigators showed that a complex-valued AR process of relatively low order (4 or 5) provides a reliable method for modeling the different forms of coherent radar clutter in an air-traffic control environment: ground clutter, rain clutter, and clutter due to migrating flocks of birds; see references 31 to 34. It is therefore ironic to find that a complex-valued AR model of somewhat similar order is also capable of modeling sea clutter under the right time scale.

4.5.3 State-Space Theory

The adoption of a *state-space model* for sea clutter is a natural choice for describing the nonstationary nonlinear dynamics responsible for its generation. Most importantly, *time* features explicitly in such a description.

The challenge in the application of a state-space model to sea clutter is basically twofold:

1. Formulation of the process (state-evolution) and measurement equations (including the respective dynamic and measurement noise processes), which are most appropriate for the physical realities of sea clutter.
2. Use of a computational procedure, which is not only efficient but also most revealing in terms of the phenomenological aspects of sea clutter.

Each of these two issues is important in its own way.

In light of the material presented in Sections 4.3 and 4.4, and contrary to conclusions reported in earlier papers [19–23], we have now come to the conclusion that sea clutter is *not* the result of deterministic chaos. By definition, the process equation of a deterministic chaotic process is noise-free. In reality, however, the process equation of sea clutter contains dynamic noise due to the fast fluctuations of the various forces, which act on the ocean surface (see the four kinds of external and internal forces, listed in Section 4.3.5). As pointed out by Heald and Stark [74], there is no physical system that is entirely free of noise, and there is no mathematical model that is an exact representation of reality. We must therefore expect dynamic noise to account for errors in the underlying model dynamics (i.e., evolution of the state with time), and we must also expect measurement noise to account for unavoidable uncertainty in matching the dependence of observables on the state.

The combined presence of dynamic noise and measurement noise in the state-space model of sea clutter has two important consequences:

1. There is unavoidable practical difficulty in disentangling the dynamic noise from the measurement noise when we try to reconstruct an invariant measure [67]. This may be the reason for why currently available algorithms for estimating chaotic invariants are incapable of discriminating between sea clutter and its stochastic surrogates in a reliable manner.
2. The delay-embedding theorem for dynamic reconstruction is formulated on the premise of a deterministic process. Although, from an experimental perspective, it is possible to account for the presence of measurement noise through a proper choice of embedding delay and embedding dimension [64], it is difficult to get around the unavoidable presence of dynamic noise in the process equation. This may explain the reason for why it is very difficult to build a predictive model for sea clutter that solves the dynamic-reconstruction problem (for varying sea state) in a reliable manner.

It is these two points that rule out deterministic chaos as a possible model for sea clutter.

4.5.4 Nonlinear Dynamical Approach Versus Classical Statistical Approach

The main focus of the classical approach, as discussed in Section 4.2, has been to model (and hopefully explain) the amplitude statistics of sea clutter. The emphasis is on point statistics, with no attention given to the temporal dimension. Some efforts have simply involved empirical fitting of distributions to the observed clutter data. Other studies have tried to provide some theoretical basis for the selection of the clutter behavior in order to make the problem mathematically tractable. For example, the assumption that we have discrete, independent scatterers permits the application of random-walk theory in developing theoretical solutions. This approach was used in the original development of the *K*-distribution [27]. However, the applicability and efficiency of the model are determined by the validity of the assumptions made in its development. The appeal of the compound *K*-distribution model is that it can be cast as the overall distribution for the product of two terms—one Rayleigh-distributed and the other Chi-distributed—which, in turn, have been found to empirically fit the two time scales of sea-clutter data in many cases. The main motivation for the development of clutter amplitude statistical models has been in their use for estimating the performance of various target-detection algorithms. However, the algorithms do not make use of the temporal properties of the clutter per se; rather, they seek to adapt the decision thresholds in response to changes in the point statistics of the clutter signal.

By contrast, the nonlinear dynamical approach, advocated in this chapter, accounts for time in an explicit manner. Moreover, the explicit need for a statistical model is avoided by using real-life data to compute the parameters of a complex AR model or state-space model of sea clutter in an on-line fashion; the complex nature of the model parameters is attributed to the in-phase and quadrature components of clutter data generated by a coherent radar. In this alternative approach, the information content of the input data is transferred directly to the model parameters evolving over time. In so doing, we are enabled to learn (a) limits on the *predictability* of sea clutter and, (b) *complexity* of its underlying dynamics, both of which have important practical implications of their own on the tracking and detection of targets in sea clutter.

4.5.5 Stochastic Chaos

In casting serious doubts in Section 4.3 on the validity of deterministic chaos as a mathematical basis for modeling sea clutter, it is important to note we are not ruling out the possibility of *stochastic chaos*, which could naturally build on the nonlinear dynamics of sea clutter.

Given a nonlinear and stochastic dynamic phenomenon, Sugihara [67] distinguishes between four different possibilities: ordinary stochastic process, stochastic chaos, ordinary deterministic process, and deterministic chaos, as depicted in Fig. 4.15. (This picture is a simplification of Fig. 2b in Sugihara's paper.) According to this figure, *stability*, or lack of it, is the feature that distinguishes stochastic chaos from an ordinary stochastic process. To elucidate this distinction, refer to the state-space model of (4.1) and (4.2), reproduced here in the form

$$\mathbf{x}_t = \mathbf{f}(\mathbf{x}_{t-1}, \mathbf{v}_{t-1}) \tag{4.17}$$

and

$$\mathbf{y}_t = \mathbf{h}(\mathbf{x}_t \mathbf{w}_t) \tag{4.18}$$

where, to simplify the exposition, we have ignored explicit dependence of the functions **f** and **h** on time t. On this basis, we may offer the following statement:

> *The underlying dynamics of a nonlinear physical phenomenon are describable by stochastic chaos, provided that the noise-free model of the phenomenon satisfies the requirements of deterministic chaos.*

The noise-free model is obtained by setting both the dynamic noise $\mathbf{v}_{t-1}$ and measurement noise $\mathbf{w}_t$ equal to zero for all time t.

The key question is: How practical is this definition for stochastic chaos? By ensuring that the observable signal-to-noise ratio is high (i.e., the measurement noise is weak), we may satisfy the condition of zero measurement noise to a good degree of approximation. Unfortunately, however, we have no control over the dynamic noise, which makes the development of a noise-free model for a physical phenomenon rather problematic.

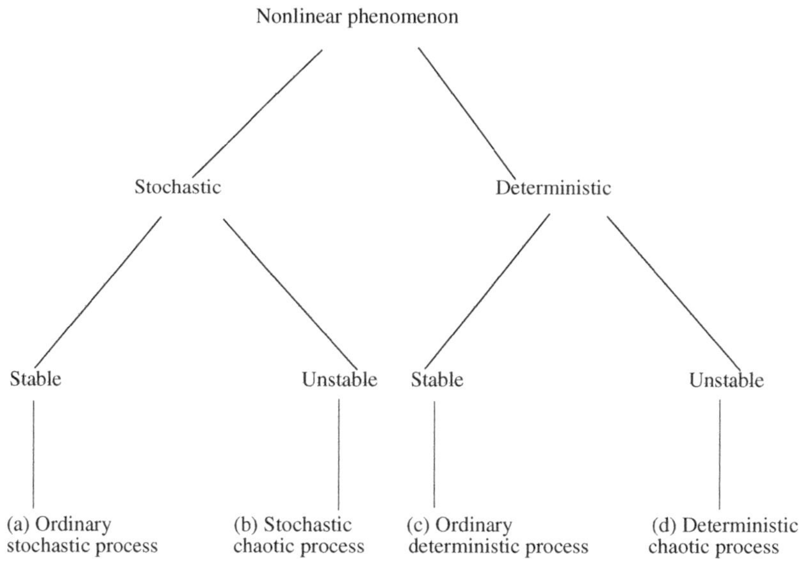

Figure 4.15 Hierarchiacal classification of nonlinear phenomenona. The picture identifies four different classes of nonlinear phenomena: (a) Ordinary stochastic process, (b) Stochastic chaotic process, (c) Ordinary deterministic process, (d) Deterministic chaotic process. The feature that distinguishes (b) stochastic chaos from (d) deterministic chaos is the presence of dynamic noise in (b) and its absence in (d); these two pheomena, however, share a common feature: instability.

To get around this difficulty, we may view the noise-free model as the "explained" part of the phenomena. For the "unexplained" part of the phenomenon, we introduce a *process noise* that accounts for the combined effects of dynamic noise and measurement noise as well as the noise-free mapping of the state onto the observable. Specifically, we combine (4.17) and (4.18) to express the observable as follows:

$$\mathbf{y}_t = \mathbf{g}(\mathbf{y}_{t-1}, \mathbf{e}_{t-1}) \tag{4.19}$$

where $\mathbf{g}$ is a new vector-valued nonlinear function, and the noise $\mathbf{e}_{t-1}$ is the *driving force* acting on the system at time $t - 1$. According to (4.19), the noise-free model is now defined by $\mathbf{y}_t = \mathbf{g}(\mathbf{y}_{t-1})$, where the new nonlinear function $\mathbf{g} = \mathbf{h}(\mathbf{f}(\mathbf{h}^{-1}(.)))$.

Some additional remarks on the *nonlinear autoregressive* model described in (4.19) are noteworthy. Specifically, in recent work described in reference 75 for the case of a scalar dynamic system that is unknown, Gunturkun and Haykin have demonstrated that given a set of observables $\{y_i\}_{i=1}^{t-1}$, it is indeed feasible to estimate the unknown driving force e_{t-1}. The estimation is based on the training of a recurrent neural network known as the echo-state network. This new on-line neural-network processing capability may provide the basis for a new strategy for the reliable detection of a weak target in sea clutter.

Referring back to the nonlinear state-space model of (4.17) and (4.18), Heald and Stark [74] show that it is possible to identify and separate the dynamic noise from the measurement noise, provided that both noise sources are additive. In particular, they use Bayesian methods to estimate the dynamic noise level and measurement noise level, both of which are shown to be accurate, provided that a good model of the dynamics is available. The noise estimates investigated in reference 74 may provide yet another way of building a nonlinear predictive model for sea clutter.

REFERENCES

1. R. E. KALMAN (1960). A new approach to linear filtering and prediction problems, *Trans. ASME J. Basic Eng.* **82**, 35–45.
2. F. A. ASCIOTI, E. BELTRAMI, T. O. CAROLL, AND C. WIRICK (1993). Is there chaos in plankton dynamics? *J. Plankton Res.* **15**(6), 603–617.
3. J. H. LEFEBVRE, D. A. GOODINGS, M. V. KAMATH, AND E. L. FALLEN (1993). Predictability of normal heart rhythms and deterministic chaos, *Chaos* **3**, 267–276.
4. A. A. TSONIS AND J. B. ELSNOR (1992). Nonlinear prediction as a way of distinguishing chaos from random fractal sequences, *Nature* **358**, 217–220.
5. N. A. GERSHENFELD AND A. S. WEIGEND (1993). The future of time series, learning and understanding, in *Time Series Prediction, Forecasting the Future and Understanding the Past* (A. S. Weigend and N. A. Gershenfeld, eds.), Addison-Wesley, Reading, MA, pp. 1–70.
6. H. D. I. ABARBANEL, R. BROWN, J. J. SIDOROWICH, AND L. S. TSIMRIN (1993). The analysis of observed chaotic data in physical systems, *Rev. Mod. Phys.* **64**, 1331–1392.
7. J. D. FARMER (1985). Sensitive dependence on parameters in nonlinear dynamics, *Phys. Rev. Lett.* **55**, 351–354.
8. J. KAPLAN AND E. YORKE (1979). Chaotic behavior of multidimensional difference equations, *Lecture Notes in Mathematics* **730**, 228–237.
9. N. H. PACKARD, J. P. CRUTCHFIELD, J. D. FARMER, AND R. S. SHAW (1980). Geometry from a time series, *Phys. Rev. Lett.* **45**, 712–716.
10. F. TAKENS (1981). Detecting strange attractors in turbulence, *Lecture Notes in Mathematics*, **898**, 366–381.
11. R. MAÑÉ (1981). On the dimension of compact invariant sets of certain nonlinear maps, *Lecture Notes in Mathematics* **898**, 230–242.
12. P. GRASSBERGER AND I. PROCACCIA (1983). Measuring the strangeness of strange attractors, *Physica D* **9**, 189–208.
13. D. RUELLE (1990). Deterministic chaos: The science and the fiction, *Proc. R. Soc. London A* **427**, 241–248.
14. A. WOLF, J. B. SWIFT, H. L. SWINNEY, AND J. A. VASTANO (1985). Determining Lyapunov exponents from a time series, *Physica D* **16**, 285–317.
15. D. S. BROOMHEAD AND G. P. KING (1986). Extracting quantitative dynamics from experimental data, *Physica D* **20**, 217–226.
16. T. SAUER, J. A. YORKE, AND M. CASDAGLI (1991). Embedology, *J. Stat. Phys.* **65**, 579–617.
17. J. J. SIDOROWICH (1992). Modelling of chaotic time series for prediction, interpolation and smoothing, in *Proceedings of IEEE ICASSP*, IV, San Francisco, pp. 121–124.
18. M. CASDAGLI (1989). Nonlinear prediction of chaotic time series, *Physica D* **35**, 335.
19. H. LEUNG AND S. HAYKIN (1990). Is there a radar clutter attractor? *Appl. Phys. Lett.* **56**, 393–395.
20. A. J. PALMER, R. A. KROPFLI, AND C. W. FAIRALL (1995). Signature of deterministic chaos in radar sea clutter and ocean surface waves, *Chaos* **6**, 613–616.
21. H. LEUNG AND T. LO (1993). Chaotic radar signal-processing over the sea, *IEEE J. Oceanic Eng.* **18**, 287–295.
22. S. HAYKIN AND X. B. LI (1995). Detection of signals in chaos, *Proc. IEEE* **83**, 94–122.
23. S. HAYKIN AND S. PUTHUSSERYPADY (1997). Chaotic dynamics of sea clutter, *Chaos* **7**(4), 777–802.

24. H. Goldstein (1951). Sea echo in propagation of short radio waves, in MIT Radiation Laboratory Series (D. E. Kerr, ed.), McGraw-Hill, New York.
25. F. A. Fay, J. Clarke, and R. S. Peters (1977). Weibull distribution applied to sea-clutter, in *Proceedings of the IEE Conference or Radar '77*, London, 101–103.
26. G. V. Trunk (1972). Radar properties of non-Rayleigh sea clutter, *IEEE Trans. Aerosp. Electron. Syst.* **8**, 196–204.
27. E. Jakeman and P. N. Pusey (1976). A model for non-Rayleigh sea echo, *IEEE Trans. Antennas Propag.* **AP-24**(6), 806–814.
28. K. D. Ward (1981). Compound representation of high resolution sea clutter, *Electronics Lett.*, **17**(16), 561–563.
29. K. D. Ward, C. J. Baker, and S. Watts (1990). Maritime surveillance radar, Part 1: Radar scattering from the ocean surface, *IEE Proc. F* **137**(2), 51–62.
30. T. Nohara and S. Haykin (1991). Canadian East Coast radar trials and the K-distribution, *IEE Proc. (London)* **F138**, 80–88.
31. S. B. Kesler (1977). Nonlinear spectral analysis of radar clutter, Ph.D. thesis, McMaster University, Hamilton, Canada.
32. S. Haykin, B. W. Currie, and S. B. Kesler (1982). Maximum-entropy spectral analysis of radar clutter, *Proc. IEEE Special Issue on Spectral Estimation* **70**(9), 953–962.
33. W. Stehwien (1989). Radar clutter classification, Ph.D. thesis, McMaster University, Hamilton, Canada.
34. S. Haykin, W. Stehwien, C. Deng, P. Weber, and R. Mann (1991). Classification of radar clutter in an air traffic control environment, *Proc. IEEE* **79**(6), 742–772.
35. M. W. Long (1983). *Radar Reflectivity of Land and Sea*, Artech House, Norwood, MA.
36. G. Neumann and W. Pierson (1966). *Principles of Physical Oceanography*, Prentice-Hall, Englewood Cliffs, NJ.
37. H. Sittrop (1977). On the sea-clutter dependency on wind speed, in *Proceedings of the IEE Conference on Radar '77*, London.
38. G. R. Valenzuela (1978). Theories for the interaction of electromagnetic waves and oceanic waves—a review, *Boundary Layer Meteorol.* **13**, 61–65.
39. C. L. Rino and H. D. Ngo (1998). Numerical simulation of low-grazing-angle ocean microwave backscatter and its relation to sea spikes, *IEEE Trans. Antennas Propag.* **46**(1), 133–141.
40. P. H. Y. Lee, J. D. Barter, B. M. Lake, and H. R. Thompson (1998). Lineshape analysis of breaking-wave Doppler spectra, *IEE Proc.—Radar, Sonar Navig.* **145**(2), 135–139.
41. H. C. Chan (1990). Radar sea-clutter at low grazing angles, *IEE Proc. F* **137**(2), 102–112.
42. J. W. Wright (1968). A new model for sea clutter, *IEEE Trans. Antennas Propag.* **AP-16**(2), 217–223.
43. E. Conte, M. Longo, and M. Lops (1991). Modelling and simulation of non-Rayleigh radar clutter, *IEE Proc. F.* **138**(2), 121–130.
44. S. Watts (1996). Cell-averaging CFAR gain in spatially correlated K-distributed clutter, *IEE Proc.—Radar, Sonar Navig.* **143**(5), 321–327.
45. R. J. A. Tough and K. D. Ward (1999). The correlation properties of gamma and other non-Gaussian processes generated by memoryless nonlinear transformation, *J. Phys. D Appl. Phys.* **32**, 3075–3084.
46. W. J. Pierson and L. Moskowitz (1964). A proposed spectral form for fully developed wind seas based on the similarity theory of S. A. Kitaigorodskii, *J. Geophys. Res.* **69**(24), 5181–5203.
47. M. A. Donelan and W. J. Pierson (1987). Radar scattering and equilibrium ranges in wind-generated waves with application to scatterometry, *J. Geophysic. Res.* **92**(5), 4971–5029.
48. D. Walker (2000). Experimentally motivated model for low grazing angle radar Doppler spectra of the sea surface, *IEE Proc.—Radar, Sonar Navig.* **147**(3), 114–120.
49. D. Walker (2001). Doppler modelling of radar sea clutter, *IEE Proc.—Radar, Sonar Navig.* **148**(2), 73–80.
50. J. R. Apel (1987). *Principles of Ocean Physics*, International Geophysics Series, Vol. 38, Academic Press, New York.
51. E. N. Lorenz (1963). Deterministic non-periodic flows, *J. Atmos. Sci.* **20**, 130–141.
52. E. Ott (1993). *Chaos in Dynamical Systems*, Cambridge University Press, Cambridge.

53. J. Theiler, S. Eubank, A. Longtin, B. Galdrikian, and J. D. Farmer (1992). Testing for nonlinearity in time series: The method of surrogate data, *Physica D* **58**.
54. A. Siegel (1956). *Non-parametric Statistics for the Behavioral Sciences*, McGraw-Hill, New York.
55. T. Schreiber and A. Schmitz (2000). Surrogate time series, *Physica D* **142**, 346.
56. J. C. Schouten, F. Takens, and C. M. van den Bleek (1994). Estimation of the dimension of a noisy attractor, *Phys. Rev. E* **50**, 1851–1861.
57. A. M. Fraser and H. L. Swinney (1986). Independent co-ordinates for strange attractors from mutual information, *Phys. Rev. A* **33**, 1134–1140.
58. A. M. Fraser (1989). Information and entropy in strange attractors, *IEEE Trans. Inform. Theory* **35**, 245–262.
59. M. B. Kennel, R. Brown, and H. D. I. Abarbanel (1991). Determining embedding dimension for phase-space reconstruction using a geometrical construction, *Phy. Rev. E* **45**, 3403–3411.
60. H. D. I. Abarbanel and M. B. Kennel (1993). Local false nearest neighbors and dynamical dimensions from observed chaotic data, *Phys. Rev. E* **47**, 3057–3068.
61. R. Brown, P. Bryant, and H. D. I. Abarbanel (1991). Computing the Lyapunov exponents of a dynamical system from observed time series, *Phys. Rev. A* **43**, 2787–2806.
62. K. Briggs (1990). An improved method for estimating Lyapunov exponents of chaotic time series, *Phys. Lett.* A **151**, 27–32.
63. C. P. Unsworth, M. R. Cowper, B. Mulgrew, and S. McLaughlin (2000). False detection of chaotic behavior in the stochastic compound K-distribution model of radar sea clutter, in *IEEE Workshop on Statistical Signal and Array Processing*, 296–300.
64. H. D. I. Abarbanel (1996). *Analysis of Observed Chaotic Data*, Springer-Verlag, New York.
65. G. S. Patel and S. Haykin (2001). Chaotic dynamics, in *Kalman Filtering and Neural Networks*, S. Haykin (ed.), Wiley, New York, pp. 83–122.
66. S. Haykin, S. Puthusserypady, and P. Yee (1998), *Dynamic Reconstruction of Sea Clutter Using Regularized RBF Networks*, ASILOMAR, Pacific Grove, CA.
67. G. Sugihara (1994). Nonlinear forecasting for the classification of natural time series, *Phil. Trans. R. Soc. Lond. Series A*, **348**, 477–495.
68. F. Gini and M. Greco (2001). Texture modeling and validation using recorded high resolution sea clutter data, *Proceedings of the 2001 IEEE Radar Conference*, Atlanta, GA, May 1–3, pp. 387–392.
69. S. Haykin (2000). *Communication Systems*, 4th edition. John Wiley & Sons.
70. M. B. Priestley (1981). *Spectral Analysis and Time Series*, Academic Press, New York.
71. T. R. Field and R. J. A. Tough (2003a). Stochastic dynamics of the scattering amplitude generating *K*-distributed noise. *J. Math. Phys.* **44**(11), 5212–5223.
72. T. R. Field and R. J. A. Tough (2003b). Diffusion processes in electromagnetic scattering generating *K*-distributed noise. *Proc. R. Soc. Lond.* 459, Series A, 2169–2193.
73. T. R. Field and S. Haykin (2007). Non-linear dynamics of sea clutter. To be submitted to *Proc. IEE Radar.*
74. J. P. M. Heald and J. Stark (2000). Estimation of noise levels for models of chaotic dynamical systems, *Phys. Rev. Lett.*, **84**(11), 2366–2369.
75. U. Gunturkun and S. Haykin (2007). Echo state networks for driving-force estimation in nonlinear dynamic systems. Submitted for publication in *IEEE Trans. Neural Networks.*

APPENDIX A SPECIFICATIONS OF THE THREE SEA-CLUTTER SETS USED IN THIS CHAPTER

The IPIX radar data used in this chapter were measured in 1993 from a clifftop near Dartmouth, Nova Scotia, at a height of 30 m above the mean sea level, facing an open view of the Atlantic Ocean of about 130°.

Table A.1 Dataset L1: Low Sea State, Sampling Frequency 2000 Hz

Date and time (UTC):	November 18, 1993, 13:13
RF frequency:	9.39 GHz
Pulselength:	200 ns
Pulse repetition frequency:	2000 Hz
Radar azimuth angle:	190°
Grazing angle:	1.4°
Range:	1200–1410 m, sampled as 8 rangebins
Range resolution:	30 m
Radar beam width:	0.9°
Width of resolution cell:	19–23 m
Significant wave height:	0.79 m
Wind:	24 km/h, coming from 340°

Table A.2 Dataset L2: Low Sea State, Sampling Frequency 1000 Hz

Date and time (UTC):	November 17, 1993, 11:57
RF frequency:	9.39 GHz
Pulselength:	200 ns
Pulse repetition frequency:	1000 Hz
Radar azimuth angle:	135°
Grazing angle:	0.4°
Range:	4200–4410 m, sampled as 14 rangebins
Range resolution:	30 m, but sampled at 15 m intervals
Radar beam width:	1°
Width of resolution cell:	73–77 m
Significant wave height:	0.84 m
Wind:	0 km/h, coming from 230°

Table A.3 Dataset H: High Sea State, Sampling Frequency 1000 Hz

Date and time (UTC):	November 17, 1993, 20:49
RF frequency:	9.39 GHz
Pulselength:	200 ns
Pulse repetition frequency:	1000 Hz
Radar azimuth angle:	190°
Grazing angle:	1.9°
Range:	900–1110 m, sampled as 14 rangebins
Range resolution:	30 m, but sampled at 15 m intervals
Radar beamwidth:	1°
Width of resolution cell:	16–19 m
Significant wave height:	1.82 m
Wind:	22 km/h (gusts to 39), coming from 220°

Chapter 5

Sea-Clutter Nonstationarity: The Influence of Long Waves[†]

Maria Greco and Fulvio Gini

5.1 INTRODUCTION

The behavior of sea-clutter is determined by the nature of the surface roughness [1, 2]. This is normally characterized in terms of two fundamental types of waves. The first type is represented by capillary waves with wavelengths (λ) on the order of centimeters or less. The second type is represented by the longer gravity waves (sea or swell) with wavelengths ranging from a few hundred meters to less than a meter [1]. In deep water, for the capillary waves we have $\lambda < 1.73\,\text{cm}$, whereas for the gravity waves the wavelength is $\lambda > 1.73\,\text{cm}$. Capillary waves are usually generated by turbulent gusts of near surface wind, and their restoring force is surface tension. On the other front, swells are produced by stable winds, and their restoring force is the force of gravity. In fact, when the wind starts to blow over a calm sea, the first waves to form are the shortest ones. As these waves build up, nonlinear interactions transfer energy to waves with larger amplitudes and longer wavelengths. This process continues until an equilibrium point is reached, where dissipation balances the tendency for wave growth. At this point, a fully developed sea exists. Since the primary transfer of energy to the sea is at very short wavelengths, a sudden cessation of wind causes the short waves to decay rapidly, while the longer waves can last several days and propagate to great distances. Consequently, at any point on the sea surface, the

[†] Part of this work has been funded by the European Office of Aerospace Research and Development (EOARD) award number FA8655-04-1-3059. The material presented in this chapter is partly based on the following papers: F. Gini and M. Greco (2002). Texture modelling, estimation, and validation using measured sea clutter data, *IEE Proc. F*, **149**(3), 115–124. M. Greco, F. Bordoni, and F. Gini (2004). X-band Sea clutter non-stationarity: The influence of long-waves, *IEEE J. Ocean Eng.* (special issue on "Non-Rayleigh Reverberation and Clutter") **29**(2), 269–283.

Adaptive Radar Signal Processing. Edited by Simon Haykin

waves are complex summations of locally generated wind waves and waves that have propagated in from other areas and different directions, resulting in complex interactions [3].

To take into account the presence of different scales of roughness in the sea surface, Wright [4] and Bass et al. [5] developed a two-scale model of the sea-surface scattering, in which the surface height is partitioned into a large-scale displacement and a small-scale displacement. For this model, it is assumed that over any patch of the surface that is large compared with small-scale lengths, but small compared with large-scale lengths, the scattering can be modeled as first-order Bragg scattering from the small-scale structure. Thus, the effect of the large-scale structure is to change the distance between a radar antenna and each point of the considered patch, which is done by tilting the surface and advecting the small-scale structure both vertically and horizontally. The effect of large-scale surface tilt is to introduce an effective amplitude modulation of the small-scale scattering [6]. Conversely, the effect of the advection is to influence the frequency content of the overall scattering.

The Bragg scattering is based on the principle that the return signals from scatterers that are half a radar wavelength apart, measured along the line of sight from the radar, reinforce each other because they are in phase [1]. The Bragg resonant length is defined by

$$\lambda_B = \frac{\lambda_0}{2\cos\theta_0} \tag{5.1}$$

where λ_0 is the wavelength of the radar signal and θ_0 is the grazing angle. The Doppler frequency shift corresponding to this wavelength is

$$f_B = \frac{C_0}{\lambda_B} = \sqrt{\frac{g}{2\pi\lambda_B} + \frac{2\pi\gamma}{\lambda_B^3}} \tag{5.2}$$

where C_0 is the intrinsic phase speed of the Bragg wave given by the wave dispersion relation, g is the acceleration of free fall, and γ is the surface tension divided by the bulk density. At microwave frequencies, the Bragg scattering is from capillary waves and the previous equation (5.2) simplifies as follows:

$$f_B \approx \sqrt{\frac{g}{2\pi\lambda_B}} \tag{5.3}$$

As a consequence, the capillary waves approaching and receding in the radar line-of-sight direction, which satisfy the condition in (5.1), give rise to two Bragg spectral lines located at $\pm f_B$, at least in the absence of other scattering phenomena as well as long waves. The magnitude of these lines depends on the azimuth look direction of the radar relative to the wind direction. With the radar pointed in the up-wind direction, the magnitude of the approaching Bragg line is larger than that for the receding line and vice versa for down-wind look direction [3]. For the cross-wind direction, the magnitudes of the approaching and receding peaks are equal.

In a real scenario, the Bragg scatterers are advected by the orbital velocity of the intermediate waves (waves with wavelengths longer than the Bragg wavelength but shorter than the radar resolution cell) and of the long-waves (waves with

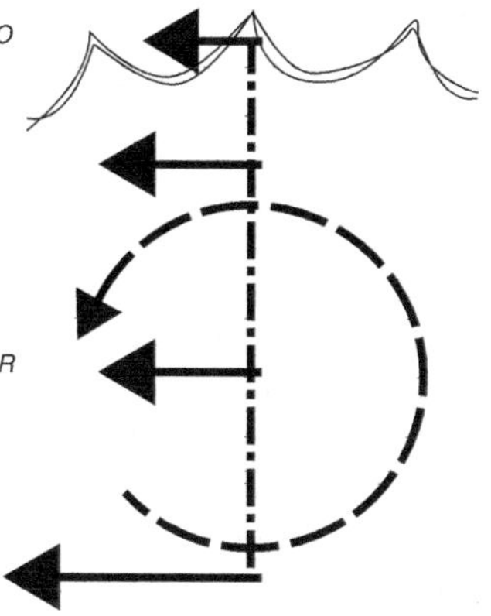

Figure 5.1 Different contributions to the surface velocity.

wavelengths longer than the radar resolution cell). The sum of the orbital velocities of the unresolved intermediate-scale waves causes a spectral broadening around the Bragg lines. For many ocean conditions, these orbital motions broaden the Bragg lines by more than their separation, causing the lines to be unresolved and generating only one Doppler peak; at X-band frequencies, this is generally the case [3, 7].

According to the Bragg theory, long-waves that are resolved by high-resolution radars may be assumed to be constant over each illuminated cell. Consequently, their effect on the Doppler spectrum is to shift the Doppler peak according to the long-waves' orbital velocity. The orbital velocities are given by the simple harmonic motion $V_0 = \pi fH = \pi H/T$, where f is the frequency of the long gravity wave of period T and H is its height from crest to trough. The contribution of the orbital velocity to the motion of the Bragg scatterers is the horizontal component—that is, $V_{OR} = V_0 \cos(2\pi ft - Kx)$, where K is the wavenumber and x is the spatial position [8]. The velocities of the scatterers in the nodes, crests, and troughs of the waves are very different. Finally, an additional Doppler shift results from any surface currents present, including wind drift. A formula often used to represent this drift is $D_w = 0.03U_w$, where U_w is the wind speed.[1] Therefore, by considering the contribution of orbital velocity, current velocity V_c, and wind drift, we can calculate the time-varying instantaneous Doppler shift as

$$f_D = \frac{2\cos\theta_0}{\lambda_0}(\pm C_0 + V_{OR} + D_w + V_c) \tag{5.4}$$

(see Fig. 5.1). Due to the periodicity of V_{OR}, f_D should be periodic as well. Actually, the Bragg scattering is not the only phenomenon determining the clutter return, particularly if breaking waves are present on the sea surface. In two recent papers [9, 10], Walker studied the development of the Doppler spectra for horizontal (HH)

[1] It is easy to observe that $V_{OR} + D_w = v_D$, following the notation of reference 1.

and vertical (VV) polarizations as breaking waves passed the radar illuminated area. Three kinds of scattering were observed: (i) Bragg scattering, present in both HH and VV polarized data, but stronger in VV data. (ii) Whitecaps scattering; the amplitude of both like-polarizations are roughly equal and are noticeably stronger than the background scatter, particularly in HH, in which case the Bragg scattering is often weak. (iii) Spikes; absent in VV data and strong only in up-wind HH data. Therefore, to see clearly the effect of the long waves on the radar returns, we should analyze datasets[2] where the Bragg scattering is dominant.

It is worth observing that in the modern statistical sea-clutter literature, the small-scale scattering structure of the two-scale model is termed *speckle* [11–15]. The variations of the local power, due to the amplitude modulation of the speckle introduced by the tilting of the small-scale structure, are modeled as a random slow-varying process referred as *texture*. According to this model, referred to as a "compound-Gaussian" distribution, the complex envelope of sea clutter can be then written as the product of fast and slow components, as shown by

$$z(n) = \sqrt{\tau(n)}x(n) \tag{5.5}$$

The fast component, $x(n)$, the *speckle*, accounts for local backscattering; it is assumed to be a stationary complex Gaussian process with zero mean and unit power. The slow component, the *texture*, $\tau(n)$, describes the underlying power level of the data; it is a non-negative real random process. Due to their different physical origins, these two components are modeled with very different correlation lengths. For X-band high-resolution sea-clutter data, the speckle correlation length was measured to be on the order of tens of milliseconds, while that of texture was on the order of seconds [16]. Because of its long correlation time, in many works on radar detection the texture is modeled as a degenerated process [17–20]; that is, it has been considered constant within each coherent processing interval (CPI), changing according to a given probability density function (pdf) from one CPI to the other. This assumption is reasonable for radar processing times that are not too large, and it gives rise, for instance, to the well-known Weibull and K models [1, 11]. As the radar processing time increases, the texture cannot be modeled as a random constant, and its variation with time should be properly taken into account. In order to predict the texture behavior, an extension to this compound-Gaussian model was proposed in reference 16, where the texture was modeled as a harmonic process, and a method for estimating the parameters of its sinusoidal components was derived and investigated. According to the compound-Gaussian model, the texture and the speckle are two independent processes: The speckle is stationary, and the texture amplitude modulates the speckle. However, as suggested by Haykin et al. [1] and as provided for by the two-scale model, the relationship between slowly and rapidly time-varying processes is much more complicated: The slowly varying swell motion modulates not only the amplitude of the speckle, but also its mean frequency (Doppler centroid) and its bandwidth.

[2] For instance, down-wind VV data.

In this chapter, first of all, evidence of the modulating effect of long waves on speckle backscattering is verified through the analysis of experimental sea-clutter data, collected in Dartmouth, Nova Scotia, at Osborne Head Gunnery Range (OHGR), with the high-resolution and low-grazing angle IPIX radar.[3] The reciprocal relationship between the long-wave evolution and the variation of the speckle spectrum shape parameters, like power (or texture), Doppler centroid, and bandwidth, is investigated by calculating cross-covariance and mutual spectrum functions. A nonstationary autoregressive (AR) process is proposed to account for and mathematically model the investigated physical phenomenon. Finally, a parametric model of the texture process is described and tested on the real-life dataset.

The effect of long waves on the overall scattering has been already considered in the literature, but generally, the analyses are theoretical; when experimental, they consider only the average spectrum—that is, the spectrum calculated on the entire dataset [3, 21, 22]. This kind of analysis can be exhaustive only when the radar resolution is low. In this case, in fact, as explained in reference 21, the low-resolution radar performs a spatial averaging over many waves and the radar cannot see the different features of the waves passing through the resolution cell during the recording process. Some evidence of the long-wave and breaking-wave amplitude- and frequency-modulation on the time-varying spectrum of real sea clutter is presented, for instance, in references 1 and 8. The novel contribution of the work reported in this chapter is to investigate this experimental evidence by means of a detailed statistical data analysis. In particular, we measure the effect of long waves on the sea-clutter amplitude and spectrum, and we propose a statistical model that takes into account the physical phenomena mentioned above.

The rest of the chapter is organized as follows. Section 5.2 describes the datasets we processed. Section 5.3 presents our statistical analysis of the sea backscatter based on the lognormal, Weibull, *K*, and generalized *K* distributions. Sections 5.4 and 5.5 develop the modeling of long-wave modulating effect on speckle backscattering through the analysis of the observed behavior of sea clutter. In particular, in Section 5.5, an autoregressive (AR) model is proposed to model the time variation of the Doppler centroid and bandwidth. In Section 5.6, a parametric model for the texture is proposed based on the periodic structure of the sea surface, and Section 5.7 concludes our findings.

5.2 RADAR AND DATA DESCRIPTION

The sea-clutter data were collected at Osborne Head Gunnery Range (OHGR) with the IPIX radar, which is an experimental X-band search radar, capable of dual-polarized and frequency-agile operation. The characteristic features of the IPIX

[3] The IPIX radar is described in Chapter 1.

Table 5.1 Data Source: OHGR Database [35]

Transmitter	Receiver	Parabolic Dish Antenna
TWT peak power: 8 kW	Coherent reception	Diameter: 2.4 m
Dual frequency simultaneous transmission: 8.9–9.4 GHz, agile	Two linear receivers, H and V on each receiver	Pencil beam width: 0.9°
H and V polarization, agile	Tuned to agile frequency	Antenna gain: 44 dB
Pulse width: 200 ns	Instantaneous dynamic range: >50 dB	Side lobes: <−30 dB
		Cross-polarization isolation: >33 dB

radar are summarized in Table 5.1 (see reference 13 for more details.) The radar site was located on a cliff facing the Atlantic Ocean, at a height of 30 meters above mean sea level with a field of view of approximately 130°. The data of the OHGR database are stored as 1-byte integers from 0 to 255. The coherent reception of both like-polarizations, HH and VV (Lpol), and cross-polarizations, HV and VH (Xpol), allows to record quadruplets of in-phase (I) and quadrature-phase (Q) values.

The characteristics and acquisition conditions of the analyzed file are summarized in Table 5.2. We processed data from all polarizations; however, here we describe in more detail the experimental results relative to the VV polarized data of the file Starea4 recorded on November 7, 1993 at 11:23 P.M. and to the HH data of the file Starea12 recorded on November 12, 1993 at 1:44 P.M. The first dataset was recorded in conditions particularly apt to reveal the long-wave modulating effects, both for the fully developed sea state and for the radar down-wind looking direction. During the time the data were collected, the Canadian Forecast Service reported that the significant wave height was of 2.23 m with an average period of 8.3 s. The wind was blowing with a speed of 4–15 km/h since 8 P.M. of November 6, coming to a relative calm state (1–3 in the Beaufort scale) after it has been blowing with strong velocity (35–45 km/h) for the previous 24 h. The wind direction was 280° from the North since 10 A.M., and the azimuth angle was fixed at 134° from the North (146° of difference, approximately down-wind). Due to these conditions in the VV data, the Bragg scattering was dominant. A pictorial view of the recording conditions of the file Starea4 is presented in Fig. 5.2. The second file was analyzed also in reference 16 to propose a parametric model of the texture process.

5.3 STATISTICAL DATA ANALYSES

The first step of our analysis was to investigate the statistical properties of the radar data. We describe here the numerical results for VV and HH data from each of the seven range cells of Starea4 file. Many distributions have been proposed in the literature to model the amplitude pdf of high-resolution, non-Gaussian clutter [11–

Table 5.2 Operative Data from OHGR Database [35]. Weather Data from the Canadian Forecast Service

Dataset Name	Starea4	Starea12	Starea4	Starea2
Date, time of the acquisition	November 7, 1993, 11:23	November 12, 1993, 13:44	November 6, 1993, 14:13	November 6, 1993, 13:40
Number of range cells	7	7	11	11
Start range	2,574 m	1,599 m	2,001 m	2,001 m
Range resolution	30 m	30 m	30 m	30 m
Range acquisition window	210 m	210 m	210 m	210 m
Pulse width	200 ns	200 ns	200 ns	200 ns
Range sample rate	10 MHz	10 MHz	5 MHz	5 MHz
Total number of sweeps	262,144	262,144	131,072	131,072
Samples for cell	131,072	131,072	131,072	131,072
Pulse repetition frequency (polarization agility)	2 kHz	2 kHz	2 kHz	2 kHz
RF frequency	9.39 GHz	9.39 GHz	9.39 GHz	9.39 GHz
Grazing angle	0.305°	0.68°	0.406°	0.455°
Azimuth angle (from N)	134°	190°	100°	211°
Wind direction (from N)	280°	202°	204°	204°
Approximate look direction	Down-wind	Up-wind	Cross-wind	Up-wind
Wind speed	7 km/h 1.94 m/s	28 to 46 km/h 7.77 to 12 m/s	40 km/h 11.11 m/s	40 km/h 11.11 m/s
Significant wave height	2.23 m	2.3 m	3.70 m	3.70 m
Significant wave period	8.3 s	4.7 s	7 s	7 s
Note sea state	Fully developed	Not fully developed	Not fully developed	Not fully developed

13, 23–25]. Here, we compare the empirical pdf of VV and HH data with lognormal (LN), Weibull (W), K, and generalized K models. Two generalized K models were considered: one with generalized Gamma distributed texture (GK) and the other with lognormal texture (LNT). The expressions of these pdf's and their moments are reported below; $R = |z(n)|$ denotes the clutter amplitude.

Lognormal Model (LN)

PDF:
$$p_R(r) = \frac{1}{r\sqrt{2\pi\sigma^2}} \exp\left(-\frac{1}{2\sigma^2}[\ln(r/\delta)]^2\right)u(r) \tag{5.6}$$

Moments:
$$\mathbf{E}\{R^n\} = \delta^n \exp(n^2\sigma^2/2), \qquad n = 1, 2, 3, \ldots \tag{5.7}$$

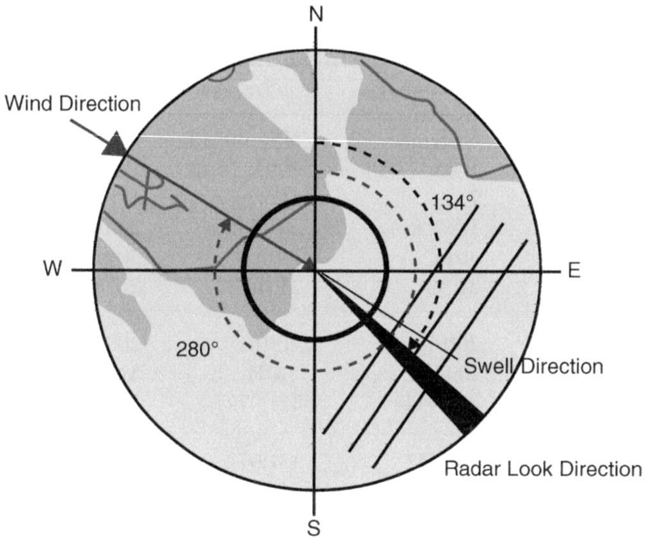

Figure 5.2 Radar and wave geometry, starea4, November 7.

where $u(\cdot)$ is the unit-step function, $\sigma > 0$ is the *shape parameter*, and $\delta > 0$ is the *scale parameter*.

Weibull Model (W)

PDF: $$p_R(r) = \frac{c}{b}\left(\frac{r}{b}\right)^{c-1} \exp\left[-(r/b)^c\right]u(r) \tag{5.8}$$

Moments: $$\mathbf{E}\{R^n\} = b^n\Gamma(n/c+1), \qquad n = 1, 2, 3, \ldots \tag{5.9}$$

where c is the *shape parameter*, and $b > 0$ is the *scale parameter*. The Rayleigh pdf is a particular case of the Weibull pdf for $c = 2$. For spiky clutter, we generally have $c \in [0.1, 1.5]$.

K-Model (K)

PDF: $$p_R(r) = \frac{\sqrt{4(\nu/\mu)}}{2^{\nu-1}\Gamma(\nu)}\left(\sqrt{\frac{4\nu}{\mu}}r\right)^{\nu} K_{\nu-1}\left(\sqrt{\frac{4\nu}{\mu}}r\right)u(r) \tag{5.10}$$

Moments: $$\mathbf{E}\{R^n\} = \left(\frac{\mu}{\nu}\right)^{n/2} \frac{\Gamma(\nu+n/2)\Gamma(n/2+1)}{\Gamma(\nu)}, \qquad n = 1, 2, 3, \ldots \tag{5.11}$$

where $\Gamma(\cdot)$ is the gamma function, $K_{\nu-1}(\cdot)$ is the modified Bessel function of the third kind of order $\nu - 1$, ν is the *shape parameter*, and μ is the *scale parameter*.[4] For spiky clutter we generally have $\nu \in [0.1, 2]$.

[4] Observe that (5.10) is equivalent to (4.7) of Chapter 4, provided that we define $c = \sqrt{\nu/u}$.

Generalized K Model with Generalized Gamma Texture (GK)

PDF:
$$p_R(r) = \frac{2br}{\Gamma(v)}\left(\frac{v}{\mu}\right)^{vb}\int_0^\infty \tau^{vb-2}\exp\left[-\frac{r^2}{\tau}-\left(\frac{v}{\mu}\tau\right)^b\right]d\tau u(r) \tag{5.12}$$

Moments:
$$\mathbf{E}\{R^n\} = \left(\frac{\mu}{v}\right)^{n/2}\frac{\Gamma(v+n/2)\Gamma(n/2+1)}{\Gamma(v)}, \qquad n = 1, 2, 3, \ldots \tag{5.13}$$

The K model is a particular case of the generalized K model. Equations (5.10) and (5.11) can be obtained from (5.12) and (5.13) by setting $b = 1$. In the limiting case, when $(v, b) \to (+\infty, 0)$ the generalized Gamma PDF reverts to the lognormal.

Generalized K Model with lognormal Texture (LNT)

PDF:
$$p_R(r) = \frac{r}{\sqrt{2\pi\sigma^2}}\int_0^\infty \frac{2}{\tau^2}\exp\left[-\frac{r^2}{\tau}-\frac{1}{2\sigma^2}[\ln(\tau/\delta)]^2\right]d\tau u(r) \tag{5.14}$$

Moments:
$$\mathbf{E}\{R^n\} = \delta^{n/2}\Gamma(n/2+1)\exp\left[\frac{1}{2}\left(\frac{n\sigma}{2}\right)^2\right], \qquad n = 1, 2, 3, \ldots \tag{5.15}$$

The characteristic parameters of all the above pdf's except for the GK-pdf are determined through the classical method of moments (MoM), by equating the first and second empirical and theoretical moments [13] (see also the recent papers [26, 27]). For example, the parameters of the K-pdf were obtained by solving the two equations

$$\mu = \mathbf{E}\{R^2\}$$
$$\frac{4v}{\pi}\left(\frac{\Gamma(v)}{\Gamma(v+1/2)}\right)^2 = \frac{\mathbf{E}\{R^2\}}{(\mathbf{E}\{R\})^2} \tag{5.16}$$

where the kth-order moment is replaced by its sample estimate:

$$\hat{\mathbf{E}}\{R^k\} = \frac{1}{N_s}\sum_{n=1}^{N_s}|z(n)|^k \to \mathbf{E}\{R^k\} \tag{5.17}$$

where N_s denotes the sample size (N_s = 131,072 for the Starea4 file). Insofar as the GK-pdf is concerned, we encountered several numerical problems with the above approach due to the flatness of the normalized moments of order 2 and 3 for the generalized K-type model as a function of the model parameters. Thus, we used a number of empirical moments greater than the number of unknowns; the parameters v and b were estimated as

$$(\hat{v},\hat{b}) = \arg\min_{(v,b)} J(v,b) = \arg\min_{(v,b)} \sum_{k=2}^{5}\left|\frac{\hat{m}_R(k)-m_R(k)}{m_R(k)}\right|^2 \tag{5.18}$$

where $m_R(k) \triangleq \mathbf{E}\{R^k\}/(\mathbf{E}\{R\})^k$ is the normalized kth-order moment (which does not depend on μ), and $\hat{m}_R(k)$ is its sample estimate. For the other distributions, we preferred to use only the two lowest useful moments, due to the higher estimation variance of the higher-order moments [13]. The absolute minimum of the functional in (5.18) was found by two successive two-dimensional searches. A first coarse grid search was carried out to prevent convergence to local minima. Then, a fine search was performed to find the absolute minimum around the one found by the previous coarse search. The fine search uses the Nelder–Mead simplex (direct search) method. Once v and b were estimated, $\hat{\mu}$ was determined from an estimate of the first-order moment, $\hat{\mathbf{E}}\{R\}$. Tables 5.3 and 5.4 report the estimates of the parameters for the data set Starea4 for VV and HH polarized data. The results of the histogram analysis for the VV data are reported in Fig. 5.3. The empirical and theoretical normalized moments, $\{m_R(k)\}_{k=1}^{6}$, are displayed in Fig. 5.4. The results show a very good fit to the data with the GK-pdf model, which belongs to the compound-Gaussian family, even if, as evident from Tables 5.3 and 5.4, the clutter backscattering is spatially heterogeneous; that is, the distribution parameters vary from cell to cell. To measure the goodness of fit, we evaluated the *root mean-square error* (RMSE) for each distribution as defined in reference 28:

$$\text{RMSE} = \frac{1}{N_p}\sum_{k=1}^{N_p} |p_R(k) - h(k)|^2 \tag{5.19}$$

where $p_R(.)$ is the generic pdf under test, $h(.)$ is the histogram, and k is the generic point of the amplitude axis in which both histogram and pdf are evaluated. We found that usually the RMSEs for the Weibull, K, GK, and LNT models are comparable and always quite small—as can be seen in Tables 5.3 and 5.4, where we report the values of RMSE for each cell and each model. In the last column (R), in both tables,

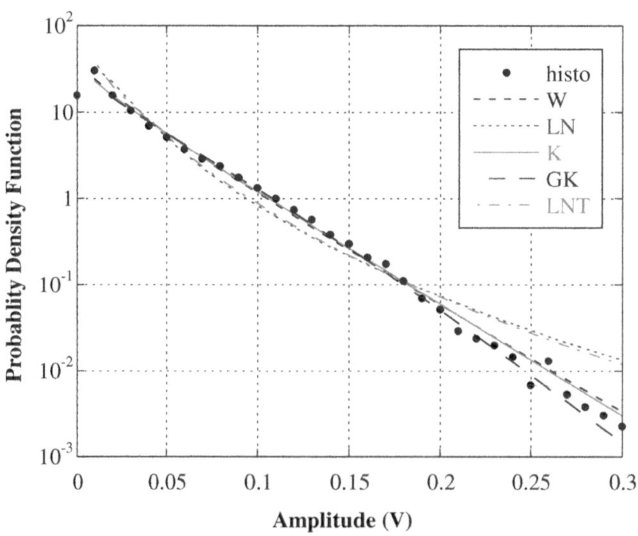

Figure 5.3 Clutter amplitude's pdf, Starea4, VV polarization, third range cell.

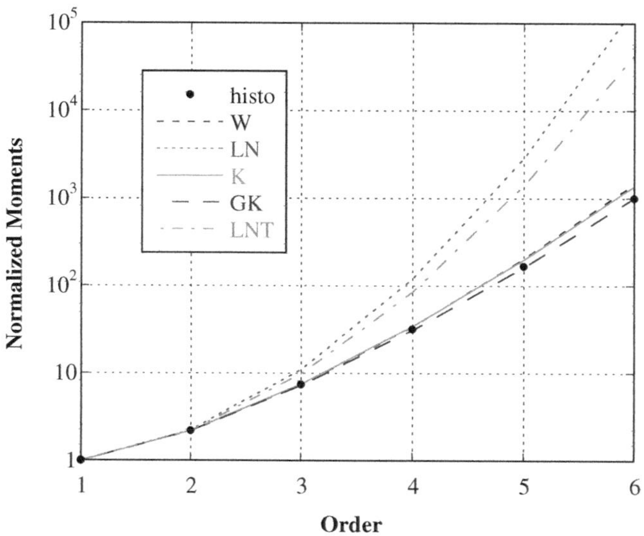

Figure 5.4 Normalized clutter moments, VV polarization, third range cell.

we report the RMSE calculated using a Rayleigh pdf with the same mean value as the data [13]. It is apparent that the data are far from the Rayleigh model. The best fitting of the moments is always obtained with the GK model. A similar statistical analysis has been carried out on the file Stareal2; here again, the analysis shows that those HH data provide a good fit to the compound-Gaussian model with *K*-distributed amplitude.

5.4 MODULATION OF LONG WAVES: HYBRID AM/FM MODEL

In this section, the physical phenomenon of the modulation induced by the long waves on the Bragg-wave speckle is analyzed. As already stated, the selection of the vertical polarization is justified because Bragg scattering was dominant (in the recording conditions of file Starea4) and the effect of long waves should be clear and evident. According to composite surface theory [1, 6], the VV backscatter has a higher averaged power in comparison with the HH backscattering. This is the case of the analyzed file, as shown in Figs. 5.5–5.7. In particular, Fig. 5.5 shows the mean HH and VV backscattered power for each range cell. The HH mean power is well below the VV mean power in all the range cells. Figure 5.6 shows the time evolution of the texture, estimated by means of the moving-window (MW) estimator, defined as

$$\hat{\tau}(l) = \frac{1}{L} \sum_{n=1+(l-1)L/2}^{(l+1)L/2} |z(n)|^2, \qquad l = 1, 2, 3, \ldots, N_B \tag{5.20}$$

Table 5.3 Estimated Parameters, File Starea4, VV Polarization

Range Cell	W			LN			K			GK				LNT			R
	$\hat{c}$	$\hat{b}$ 10^{-2}	RMSE 10^{-4}	$\hat{\delta}$ 10^{-4}	$\hat{\sigma}^2$	RMSE 10^{-4}	$\hat{V}$	$\hat{\mu}$ 10^{-4}	RMSE 10^{-4}	$\hat{V}$	$\hat{\mu}$ 10^{-4}	$\hat{b}$	RMSE 10^{-4}	$\hat{\delta}$ 10^{-4}	$\hat{\sigma}^2$	RMSE 10^{-4}	RMSE 10^{-4}
1st	0.91	2.57	1.20	1.00	0.87	78.5	0.39	8.07	3.06	0.99	5.14	1.16	3.65	2.63	2.24	57.3	961
2nd	0.96	2.43	1.69	85.1	0.86	27.5	0.45	6.33	2.13	0.45	1.40	1.99	2.23	6.33	1.93	18.0	502
3rd	0.93	2.79	21.6	1.15	0.89	219	0.39	9.17	13.0	0.26	34.9	2.92	3.50	3.09	2.18	175	1580
4th	0.82	2.98	16.1	1.35	0.96	79.2	0.30	13.6	54.2	0.86	9.60	1.08	3.80	3.53	2.70	52.7	1560
5th	0.84	3.25	7.16	1.66	0.94	200	0.31	15.7	16.5	0.25	60.5	2.28	19.0	4.19	2.64	151	2060
6th	0.84	3.08	7.36	1.48	0.94	219	0.31	14.1	7.15	0.26	52.3	2.20	8.39	3.78	2.64	170	2100
7th	0.80	3.44	30.8	1.90	0.97	332	0.28	19.4	11.4	0.18	94.0	2.75	15.2	4.75	2.82	269	2360

Table 5.4 Estimated Parameters, File Starea4, HH Polarization

Range Cell	W			LN			K			GK				LNT			R
	$\hat{c}$	$\hat{b}$ 10^{-3}	RMSE 10^{-6}	$\hat{\delta}$ 10^{-6}	$\hat{\sigma}^2$	RMSE 10^{-6}	$\hat{V}$	$\hat{\mu}$ 10^{-5}	RMSE 10^{-6}	$\hat{V}$	$\hat{\mu}$ 10^{-7}	$\hat{b}$	RMSE 10^{-6}	$\hat{\delta}$ 10^{-5}	$\hat{\sigma}^2$	RMSE 10^{-6}	RMSE 10^{-6}
1st	0.81	5.54	46.0	2.81	0.97	11.5	0.27	5.00	55.8	4.30	0.18	0.45	14.4	1.19	2.88	14.9	62.2
2nd	1.05	5.84	$<10^{-6}$	3.31	0.79	0.33	0.60	3.00	$<10^{-6}$	1.12	<0.001	0.42	0.04	1.40	1.56	0.10	10
3rd	0.96	6.22	18.3	3.71	0.86	5.51	0.46	4.20	18.5	58.2	<0.001	0.16	6.03	1.55	1.98	7.97	18.7
4th	0.55	4.60	817	2.45	1.24	304	0.12	15	1070	0.84	188	0.58	725	1.06	5.30	221	236
5th	0.76	6.90	104	4.89	1.01	25.5	0.26	9.3	188	2.72	6.09	0.54	25.2	1.96	3.12	29.9	386
6th	0.88	7.01	53.6	4.90	0.91	13.7	0.36	6.4	60.8	1.85	59.8	0.78	26.8	1.97	2.35	15.5	72.1
7th	0.73	7.30	139	5.49	1.04	59.2	0.24	116	240	1.63	65.4	0.64	74.7	2.14	[illegible]	68.2	661

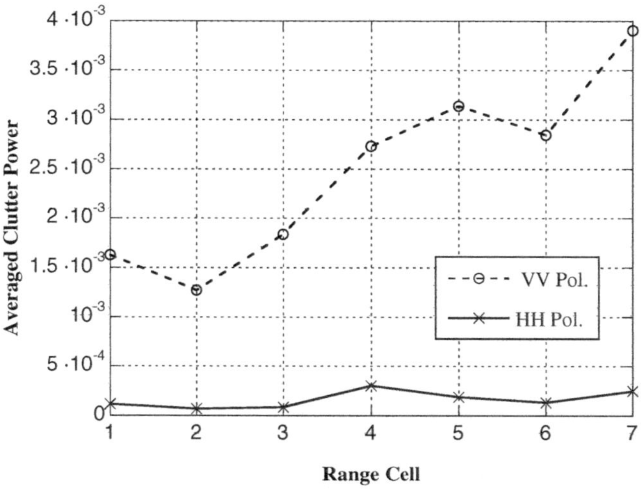

Figure 5.5 Mean backscattered power for VV and HH polarization versus the corresponding range cell. Dataset: Starea4 of November 7.

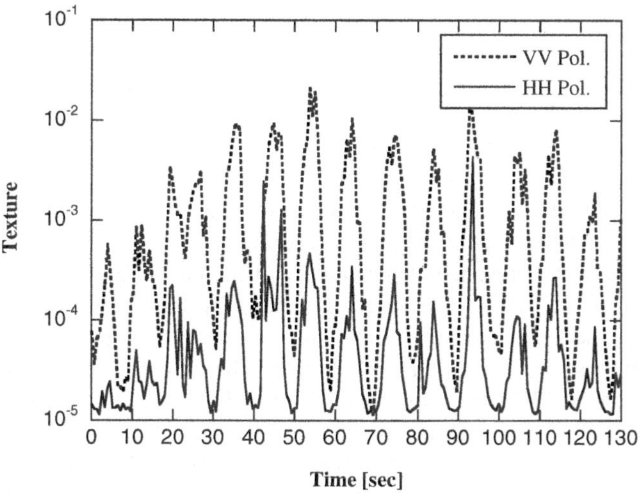

Figure 5.6 HH and VV polarized backscattering mean texture versus the observation time interval. Dataset: Starea4 of November 7, range cell 3.

where $z(n)$ is the nth sample of the complex envelope of the data, $N_B = 2047$ is the number of bursts in which the entire sequence of the range cell data has been divided, and $L = 128$ is the number of samples in the lth burst (adjacent bursts have an overlap of 50%). We assumed and verified that the texture is constant within each short burst of 0.128 s. The almost-periodic behavior of the texture is evident for both polarizations. Finally, Fig. 5.7 plots the ratio of the HH and VV textures

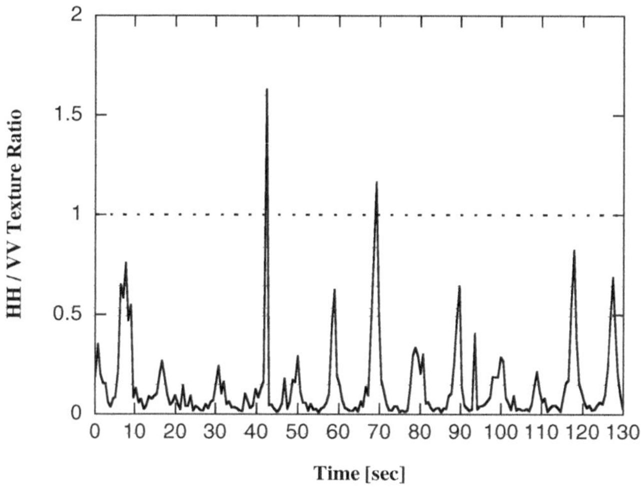

Figure 5.7 Polarization ratio versus the observation time interval. Dataset: Starea4 of November 7, range cell 3.

versus the time observation interval; the spike occurrence is clearly evident in the figure. When the spikes are absent, the HH–VV ratio is always below unity [10]. In the third range cell, we have only two spikes, one located at 40 s and the other one at 68 s; it is therefore apparent that Bragg scattering is dominant.[5] As evidenced by the statistical analysis, the data of the Starea4 file has a spiky behavior; in particular, the pdf exhibits long tails in both like- and cross-polarizations, even if it seems that Bragg scattering is dominant with respect to spike scattering.

In the frequency domain, the presence of periodicity in the power, and then of an amplitude-modulating effect of the swells, is evidenced by the time-varying spectrogram. In Fig. 5.8a, the spectrogram occupies the frequency range –200 to 200 Hz, where we have most of the sea clutter contribution. The time-Doppler plots were obtained by computing the FFT over a sliding window of 512 samples (0.512 s), weighted by a Hanning window and with an overlap of 50%. The 512-point FFT ($N_F = 512$) provides the spectrogram in the frequency range –500 to 500 Hz. It is apparent that the mean texture and spectrogram illustrate a common periodic trend with a period of about 10 s, which is comparable with those of the swells after projection on the radar line of sight. However, the effect of long waves is not limited to amplitude modulation. Rather, it involves a periodic change in the shape, bandwidth, and maximum of the time-varying spectrum. This aspect of the sea-clutter behavior is better revealed by the normalized spectrogram reported in Fig. 5.8b (the spectrogram in Fig. 5.8a was not normalized). The entire sequence of data has been

[5] Note that the term *spike* has two different meanings in statistics and scattering theory. Generally, in statistical analysis, we say that the clutter is spiky when in its time history, it presents some amplitude peaks, more frequently and a more pronounced manner than in the Gaussian case.

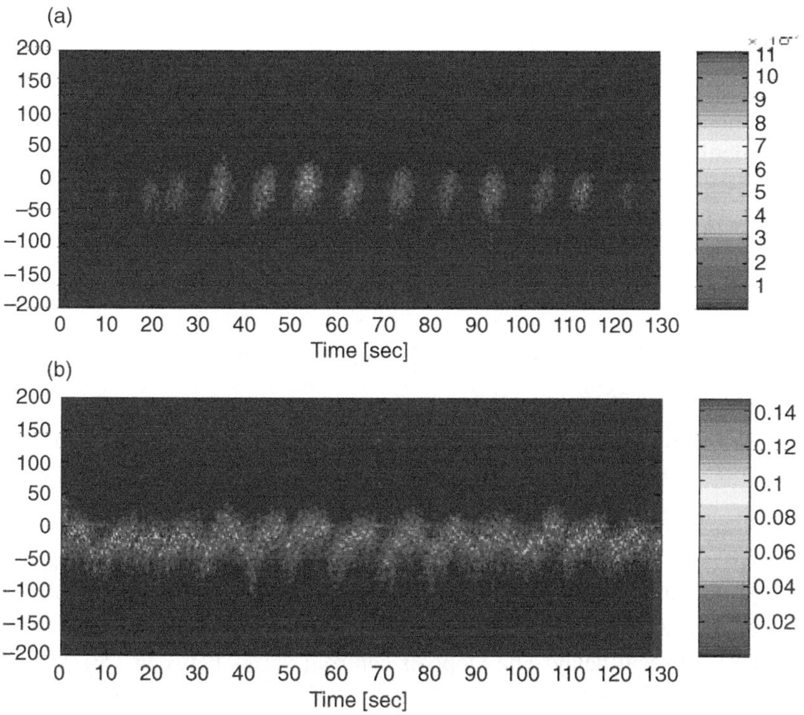

Figure 5.8 (a, b) VV spectrogram and normalized spectrogram. Dataset: Starea4 of November 7, range cell 3. See color insert.

again divided into $N_B = 2N_s/L_b - 1 = 511$ bursts of $L_b = 512$ samples each, with an overlap between adjacent bursts of 50%. Individual spectra were calculated using a 512-point FFT and Hanning windowing, and then normalized with respect to the local-clutter power. The nonstationarity of the sea clutter related to the periodic spectral variations is evident. These variations correspond to the changes in the frequency extent and the Doppler peak due to the frequency modulation, which is induced by the long-waves on the Bragg-wave scattering.

In order to measure the temporal variation of the backscattered Doppler spectra, we measured the Doppler centroid of the clutter spectrum and the root mean-square bandwidth, defined respectively as

$$f_C = \frac{\int_{-\infty}^{\infty} f S(f)\,df}{\int_{-\infty}^{\infty} S(f)\,df} \tag{5.21}$$

and

$$B_W = \sqrt{\frac{\int_{-\infty}^{\infty} (f - f_C)^2\, S(f)\,df}{\int_{-\infty}^{\infty} S(f)\,df}} \tag{5.22}$$

where f is the frequency, and $S(f)$ is the clutter power spectral density (PSD). We investigated the behavior of the Doppler centroid and not of the PSD peak because it is generally more representative of the short-time spectral change, especially in the case of a nonsymmetric spectral shape. The time-varying centroid and bandwidth were calculated respectively as follows:

$$\hat{f}_C(l) = \frac{1}{Q} \sum_{n=-N_F/2}^{N_F/2-1} f(n) P_l(n), \qquad l = 1, 2, \ldots, N_B \tag{5.23}$$

and

$$\hat{B}_W(l) = \sqrt{\frac{1}{Q} \sum_{n=-N_F/2}^{N_F/2-1} \left[f(n) - \hat{f}_C(l)\right]^2 P_l(n)}, \qquad l = 1, 2, \ldots, N_B \tag{5.24}$$

where $f(n) = 1000n/N_F$ hertz is the digital frequency, $P_l(n)$ is the periodogram of the lth data burst (composed of 512 samples) corresponding to $f(n)$, computed on $N_F = 512$ points and $Q = \sum_{n=-N_F/2}^{N_F/2-1} P_l(n)$.

Figure 5.9 shows, on arbitrary scale and units for ease of presentation, the texture temporal evolution, the Doppler centroid, and bandwidth, obtained by processing the data from the third range cell. They exhibit a common periodic trend with their own reciprocal delay. The Doppler centroid varies between −50 Hz and 5 Hz, and the bandwidth varies between 25 Hz and 225 Hz. We observe that the maxima in the texture values are coincident with the minima of the bandwidth, and vice versa. The spectral enlargement could be due to the modulation induced by the long waves or due to the presence of thermal noise. The effect of this noise in the spectral and temporal features of the received radar returns may be quite different,

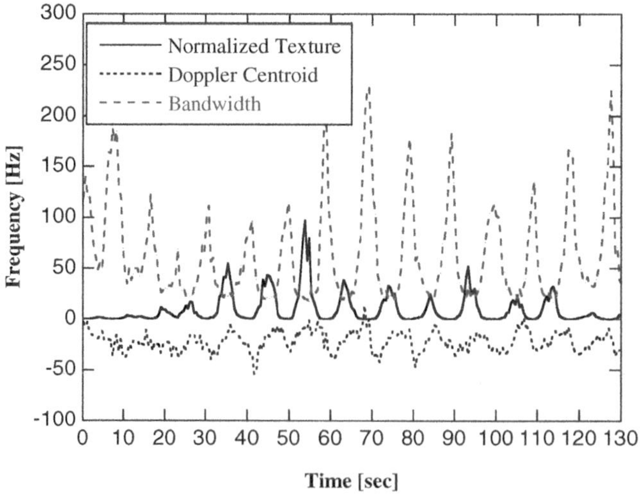

Figure 5.9 Time evolution of VV polarized backscattering mean texture, Doppler centroid, and backwidth. Dataset: Starea4 of November 7, range cell 3, VV polarization.

depending on the clutter-to-noise power ratio (CNR). According to our estimates, the CNR ranges from about 80 dB when the texture is maximum down to about −5 dB when the texture is minimum. In the latter case, the contribution of noise to the radar return is strong. The Doppler centroid and the bandwidth follow the texture behavior, but with small delays of 2–3 s and 5 s, respectively—that is, about 1/4 and 1/2 of the swell period.

In order to quantify the relationship between texture, centroid, and bandwidth, we calculated their cross-covariance functions (CCF). The CCF between the time-varying texture and Doppler centroid was calculated as

$$C_{\tau C}(m) = \frac{\sum_{l=1}^{N_B-|m|} [\hat{\tau}(l) - \hat{\eta}_\tau] \cdot \left[\hat{f}_C(l+m) - \hat{\eta}_C\right]}{\sum_{l=1}^{N_B} [\hat{\tau}(l) - \hat{\eta}_\tau]^2 \sum_{l=1}^{N_B} \left[\hat{f}_C(l) - \hat{\eta}_C\right]^2}, \quad m = -(N_B - 1), \ldots, N_B \tag{5.25}$$

and the CCF between texture and bandwidth was calculated as

$$C_{\tau W}(m) = \frac{\sum_{l=1}^{N_B-|m|} [\hat{\tau}(l) - \hat{\eta}_\tau] \cdot \left[\hat{B}_W(l+m) - \hat{\eta}_w\right]}{\sqrt{\sum_{l=1}^{N_B} [\hat{\tau}(l) - \hat{\eta}_\tau]^2 \sum_{l=1}^{N_B} \left[\hat{B}_W(l) - \hat{\eta}_W\right]^2}}, \quad m = -(N_B - 1), \ldots, N_B \tag{5.26}$$

where, $\hat{\tau}(l)$, $\hat{f}_C(l)$, and $\hat{B}_W(l)$ are, respectively, the estimated texture, Doppler centroid, and bandwidth; and

$$\hat{\eta}_\tau = \frac{1}{N_B} \sum_{l=1}^{N_B} \hat{\tau}(l), \quad \hat{\eta}_C = \frac{1}{N_B} \sum_{l=1}^{N_B} \hat{f}_C(l), \quad \hat{\eta}_w = \frac{1}{N_B} \sum_{l=1}^{N_B} \hat{B}_W(l) \tag{5.27}$$

are their respective sample means. Figures 5.10a and 5.10b show the plots of $C_{\tau C}(m)$ and $C_{\tau W}(m)$, respectively. They show a behavior very similar to the cross-covariance functions between two truncated sinusoids with a common period of about 10 s separated by a lag of about 2.5 s and 5 s, respectively.

In order to retrieve the actual frequency values of the common sinusoidal components, we calculated the absolute value of the mutual PSDs—that is, the absolute value of the Fourier transform of $C_{\tau C}(m)$ and $C_{\tau W}(m)$. The results are shown in Figs. 5.11a and 5.11b. The information on the frequency "similarity" of texture and centroid provided by these mutual PSDs is analogous to that contained in the *modulation transfer function* (MTF), which is well known in the geophysics literature (see reference 6 and references therein). The peaks in the mutual PSD correspond to the frequencies common to the two analyzed signals. This means, for instance, that the frequency component at $f_{LW} = 0.1$ Hz is present in both texture and Doppler centroid signals and in both texture and bandwidth signals. This fact is confirmed by the normalized Fourier spectrum of texture, Doppler centroid, and bandwidth (calculated after sample-mean removal). The curves are presented in Figs 5.12a–5.12c;

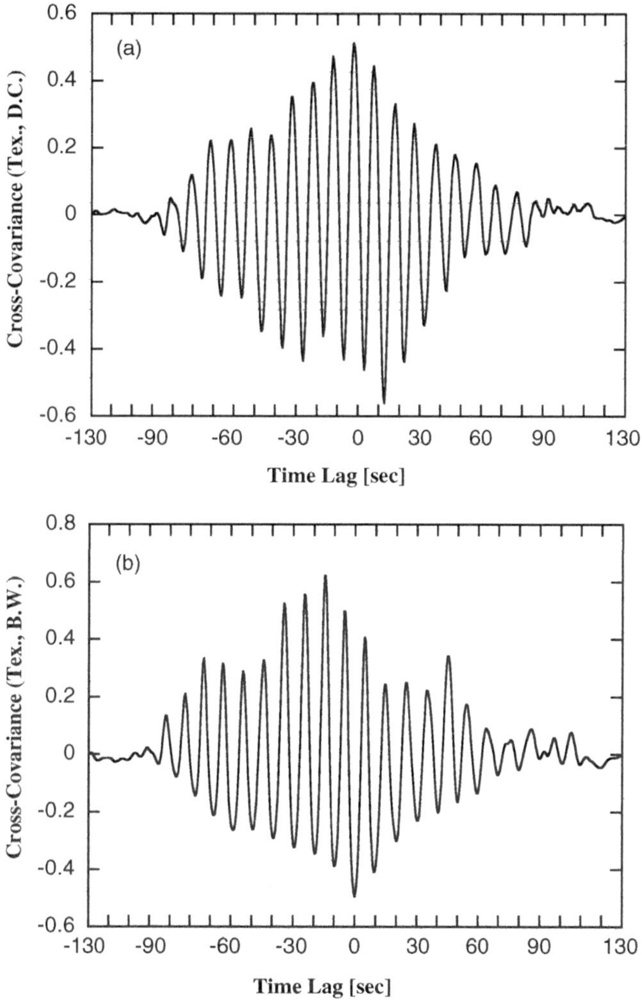

Figure 5.10 (a, b) Cross-covariance between mean texture and Doppler centroid and between mean texture and bandwidth. Dataset: Starea4 of November 7, range cell 3, VFV polarization.

they confirm the hypothesis of a near line-shape Doppler spectrum with components located in correspondence with the long-wave frequency $f_{LW} = 0.1\,\text{Hz}$. Moreover, secondary components are present close to $f \approx 0$ and, especially for texture and bandwidth spectra, at $f = 2f_{LW}$. These results suggest that, at least when Bragg scattering is dominant, the long waves modulate the speckle scattering in amplitude, causing the variation of the local power, and in frequency, causing a periodic variation of the spectrum centroid and bandwidth.

The specific conditions present when the Starea4 (November 7) data were collected make them particularly apt to show the long-wave modulating effects. The

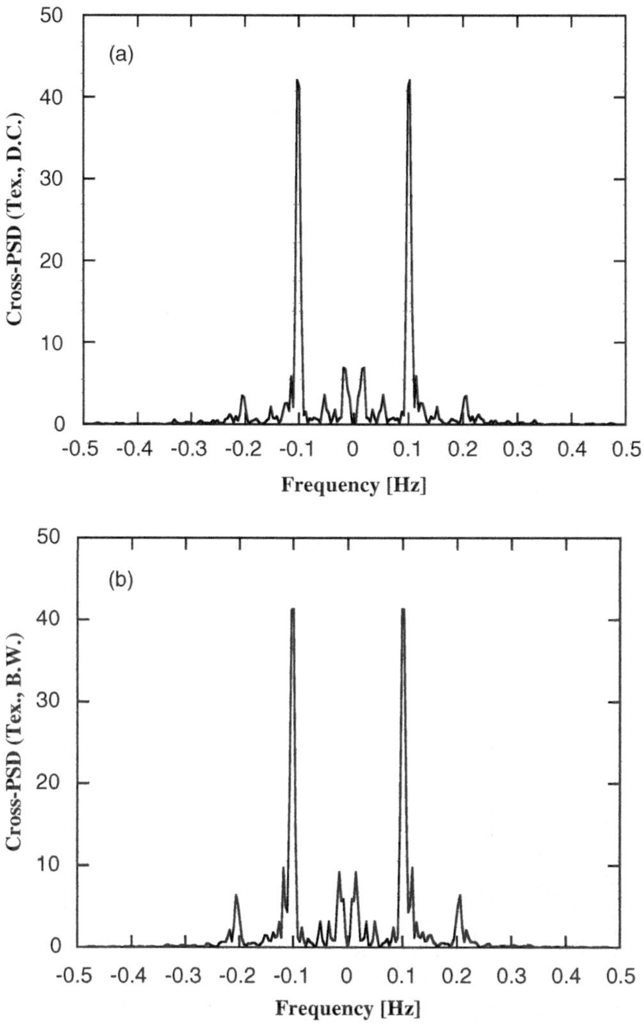

Figure 5.11 (a, b) Absolute value of the normalized cross-spectrum between mean texture and Doppler centroid and between mean texture and bandwidth, respectively. Dataset: Starea4 of November 7, range cell 3, VV polarization.

same cannot be said for the other analyzed files such as Starea2 and Starea4, recorded on November 6; Starea12, recorded on November 12; the first and the third, in near up-wind condition; and the second, in cross-wind condition. In fact, based on the Canadian Forecast Service reports, it is reasonable to presume that the sea was not in a fully developed state (see Section 5.2). When the sea is not in a fully developed state and the radar look direction is not down-wind so that the sensor can see the breaking waves, the assumption of predominance of Bragg scattering does not hold any more, and the modulating effect is therefore less evident. However,

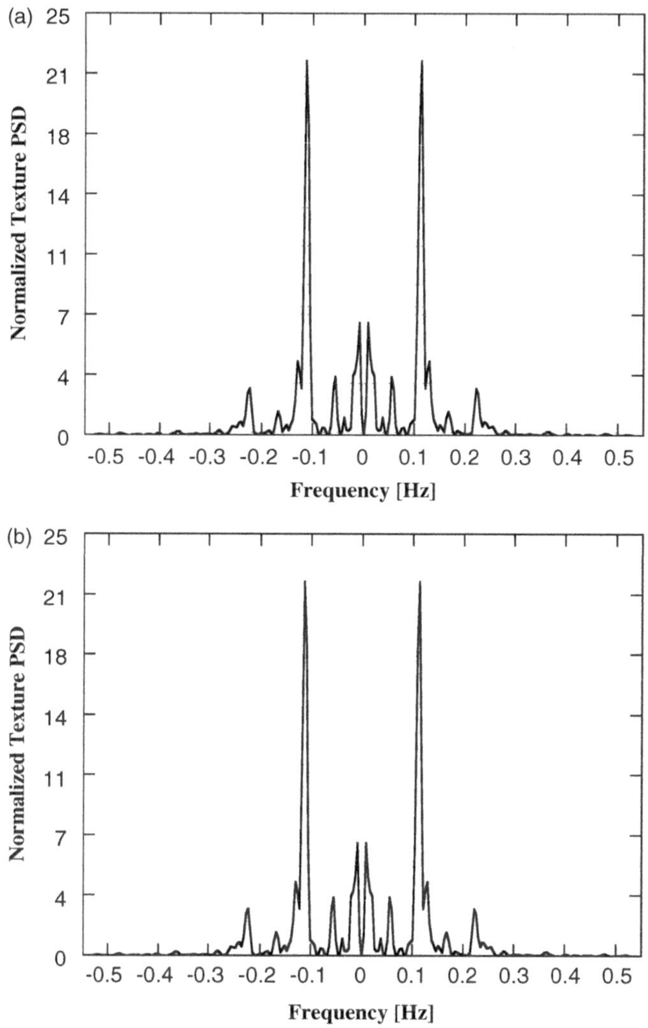

Figure 5.12 (a–c) Normalized power spectral density of the mean texture, Doppler centroid, and bandwidth, respectively. Dataset: Starea4 of November 7, range cell 3, VV polarization. Parts (a) and (b) of the figure are displayed here; part (c) is displayed on the next page.

except for the data collected in cross-wind conditions, Starea4 of November 6, the time evolution of texture, Doppler centroid, and bandwidth still exhibit marked common behaviors. These observations are also confirmed by the cross-covariance functions and normalized Fourier spectrum. In fact, common spectral lines are located in correspondence of frequency comparable with that of the long waves, though less defined as compared with the results first reported in reference 29.

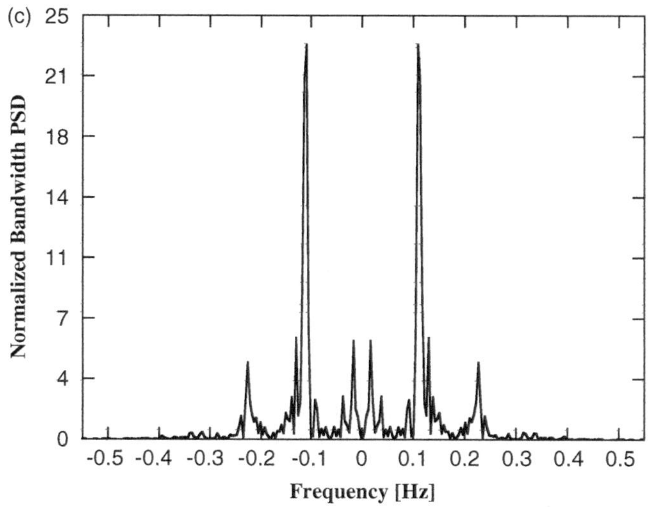

Figure 5.12 *(continued)*

5.5 NONSTATIONARY AR MODEL

There are two goals related to clutter modeling. The first is to provide further insight into the physical and electromagnetic factors that play a role in forming the clutter signal. The second is to produce a mathematical model, physically grounded, with which the clutter signal can be generated and processed to test detection algorithms. In this section, we suggest the use of AR modeling to describe the physical phenomenon of the long-wave modulation analyzed in Section 5.4. As demonstrated by others, the autoregressive models can fit both Gaussian and non-Gaussian (as in our case) processes [30, Chapter 4].

A model's ability to describe a physical phenomenon of interest is not the main issue; rather, its mathematical tractability is very important. We found that an AR model of third order, AR(3), satisfies both of these issues. With regards to tractability, the AR(3) model is defined by seven real parameters, the complex value of its three poles, and the final prediction error. Moreover, an AR process can be simply generated by properly filtering white noise. AR models of order p ranging from 1 to 16 were investigated. The normalized periodogram was obtained by averaging the individual normalized Fourier spectra related to single sliding bursts. The entire sequence of data has been divided into $N_B = 2N_s/L - 1 = 511$ bursts of $L = 512$ samples each, with an overlap between adjacent bursts of 50%. Next, the individual power spectra were calculated using a 512-point FFT with a Hanning window, and then normalized with respect to the local clutter power. With regard to the AR spectrum, it was obtained by processing all the range cell samples via Yule–Walker's method [31]. In Fig. 5.13, the results so obtained are shown. It seems that the AR(3) accurately describes the basic shape features such as Doppler centroid and bandwidth. This is confirmed by Figs. 5.14a and 5.14b, which represent the time evolution

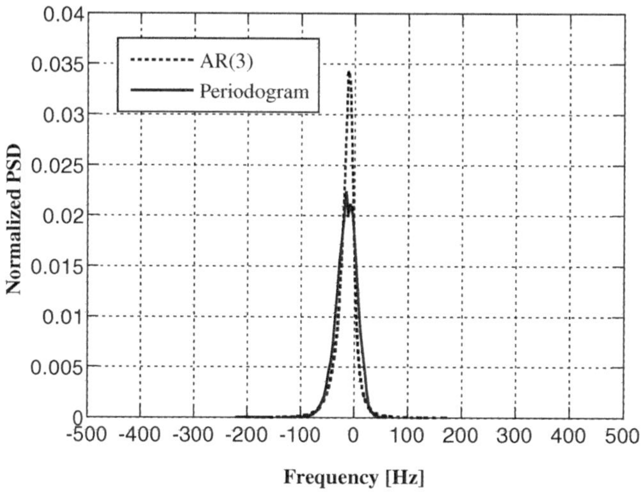

Figure 5.13 Normalized periodogram and AR(3) PSD. Dataset: Starea4 of November 7, range cell 3.

of these quantities, estimated both by the spectrogram and by the AR method. In particular, the estimated AR centroid is given by

$$\hat{f}_{C,AR}(l) = \frac{1}{Q_{AR}} \sum_{n=-N_F/2}^{(N_F/2)-1} f(n) P_{AR,l}(n), \quad l = 1, 2, \ldots, N_B \tag{5.28}$$

while the estimated AR bandwidth is

$$\hat{B}_W^{AR}(l) = \sqrt{\frac{1}{Q_{AR}} \sum_{n=-N_F/2}^{N_F/2-1} \left(f(n) - \hat{f}_C^{AR}\right) P_{AR,l}(n)}, \quad l = 1, 2, \ldots, N_B \tag{5.29}$$

where $f(n) = 1000n/N_F$ hertz is the digital frequency, $P_{AR,l}$ is the AR spectrum computed by processing the 512 samples of the lth data burst via Yule–Walker's method on $N_F = 512$ points, and $Q_{AR} = \sum_{n=-N_F/2}^{(N_F/2)-1} P_{AR,l}(n)$. Figures 5.15a and 5.15b show the power spectrum of the AR(3) centroid and AR(3) bandwidth; both AR plots preserve the spectral components of the corresponding quantities obtained by the Fourier analysis. Also, the time delay with respect to texture is preserved, as shown by the cross-covariance in Figs. 5.16a and 5.16b. Therefore, our analysis suggests that AR(3) model is indeed capable of modeling the basic effects of long-wave modulation on the capillary backscattering.

It is also of interest to investigate how the AR model parameters themselves vary with time. Figures 5.17a–c show the time evolution of the two dominant AR poles and final prediction error, together with texture. The elements of correlation are evident. The dominant pole has almost constant unitary modulus, and an instantaneous frequency that changes with the same period of the texture. Also, the second

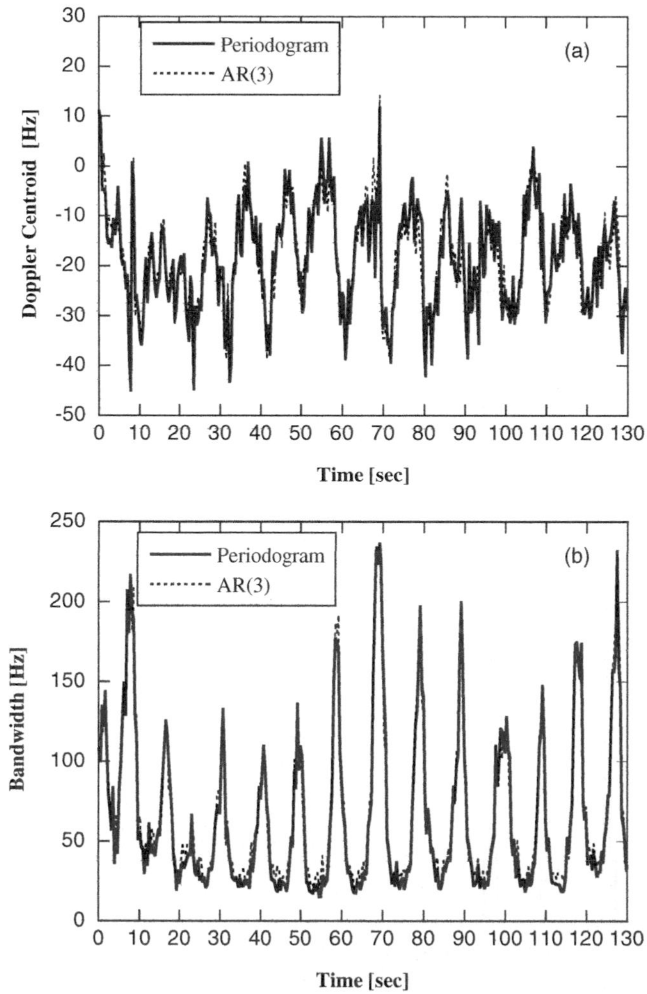

Figure 5.14 (a, b) Time evolution of the Doppler centroid and bandwidth, obtained from the Fourier and AR(3) estimated spectra. Dataset: Starea4 of November 7, range cell 3, VV polarization.

pole follows the same periodic trend, being almost constant when the signal is strongly reflected by the crest of a wave and noisy otherwise. The third pole is dominated by noise, while the final prediction error follows the texture trend.

5.6 PARAMETRIC ANALYSIS OF TEXTURE PROCESS

After analyzing the relations between texture, Doppler centroid, and bandwidth of the sea clutter, in this section we describe in detail a parametric modeling and analysis of the texture process, performed on the HH data of the file Starea12. The

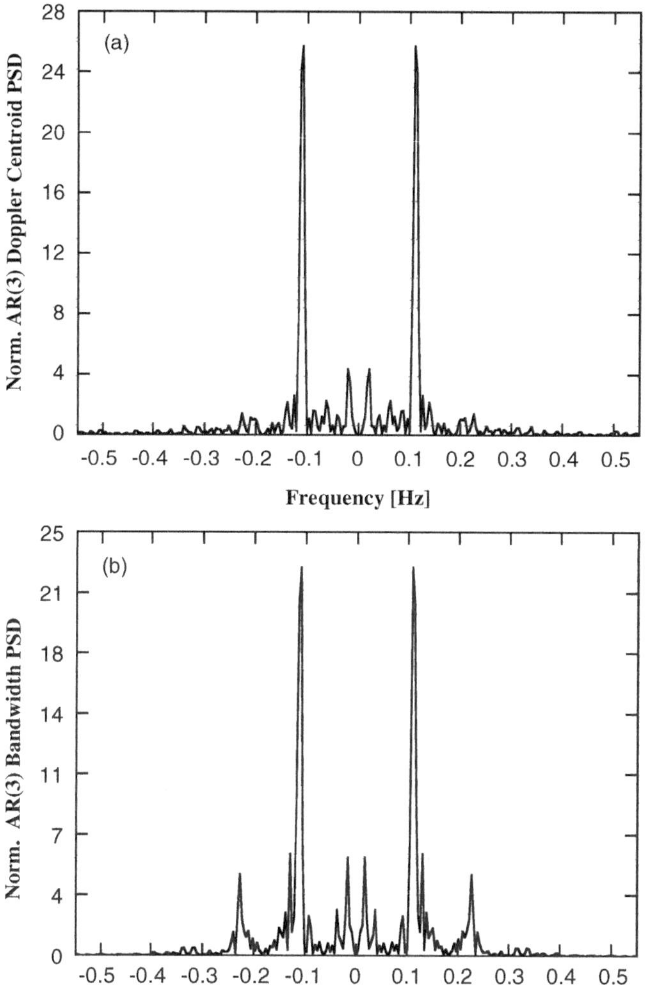

Figure 5.15 (a, b) Normalized Fourier spectrum of the AR(3) Doppler centroid, and AR(3) bandwidth. Dataset: Starea4 of November 7, range cell 3.

proposed model is based on the periodic behavior of the texture as shown in the previous section and on the cyclostationarity of the sea clutter, already theorized by Conte and Longo in [12] and empirically revealed by Haykin and Thompson in reference 32 and by Gini and Greco in reference 16; see also Chapter 3, which is an update version of reference 32. To investigate the presence of cyclostationarity in our data, it is useful to work with the intensity of the radar echo—that is, the modulus-squared of the complex envelope samples—instead of the complex envelope samples. For compound Gaussian-distributed clutter, the intensity is given by $I(n) = |z(n)|^2 = \tau(n)|x(n)|^2$.

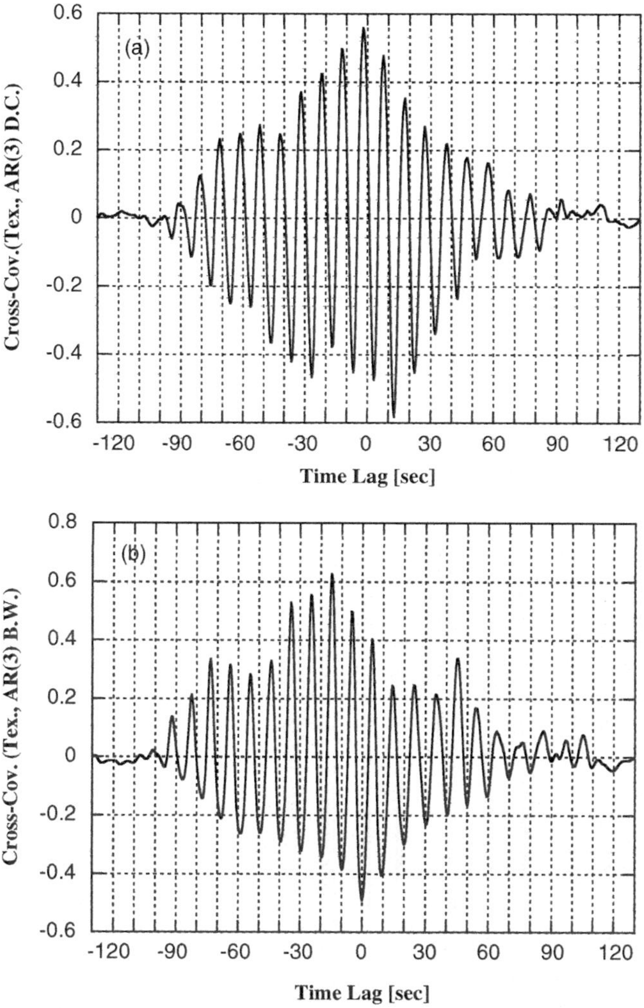

Figure 5.16 (a, b) Cross-covariance between mean texture and the AR(3) Doppler centroid and between mean texture and the AR(3) bandwidth. Dataset: Starea4 of November 7, range cell 3.

First of all, we briefly introduce some concepts of cyclic statistics (for more details see references 16, 33, and 34). The time-varying mean of a general non-stationary complex process $z(n)$ is defined as $m_{1Z}(n) \triangleq \mathbf{E}\{x(n)\}$. The process is referred to as being first-order cyclostationary (CS) if its time-varying mean is (almost) periodic[6] [33, 34]. The coefficients in the generalized Fourier series expansion of m_{1Z} are called the *cyclic mean*:

[6] Cyclostationary processes are those whose statistics are periodic functions of time. Here, "almost" refers to the fact that these periods may not be integers as required by a discrete-time framework.

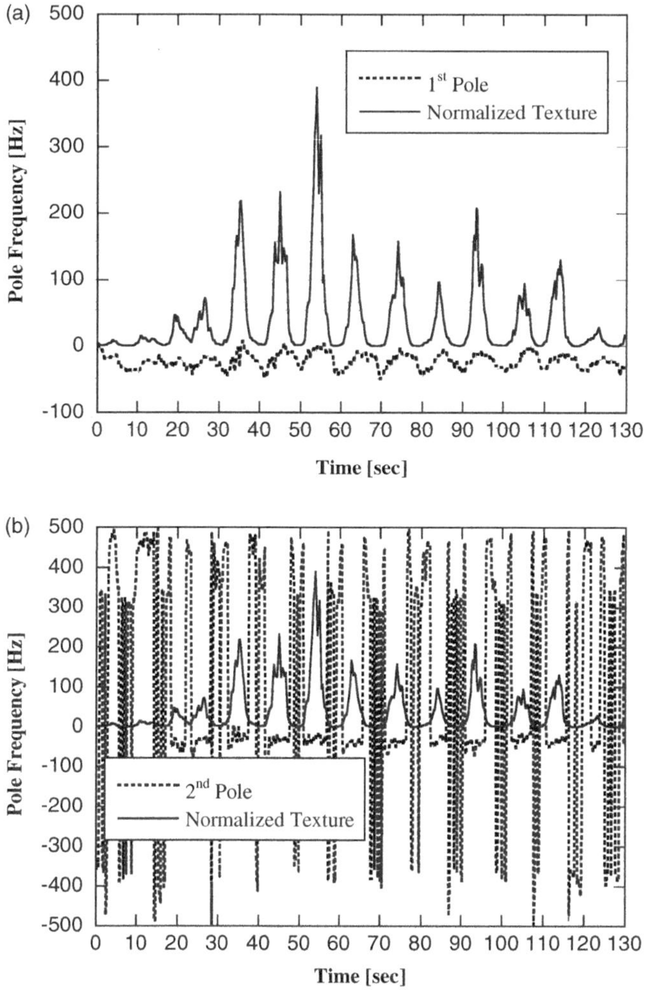

Figure 5.17 (a–c) AR(3) pole frequencies and final prediction error, compared with the normalized texture versus the observation interval. Dataset: Starea4 of November 7, range cell 3, VV polarization. Parts (a) and (b) of the figure are displayed here; part (c) is displayed on the next page.

$$M_{1Z}(\alpha) \triangleq \lim_{N\to\infty} \frac{1}{N} \sum_{n=0}^{N-1} m_{1Z}(n) e^{-j\alpha n} \tag{5.30}$$

where α is the so-called *cycle frequency*. If $z(n)$ is first-order stationary, then $M_{1Z}(\alpha) = 0$ for $\alpha \neq 0$. The event $M_{1Z}(\alpha) \neq 0$ for nonzero cycles is revealing of data nonstationarity. For the intensity of the compound Gaussian process, the cycle mean is given by

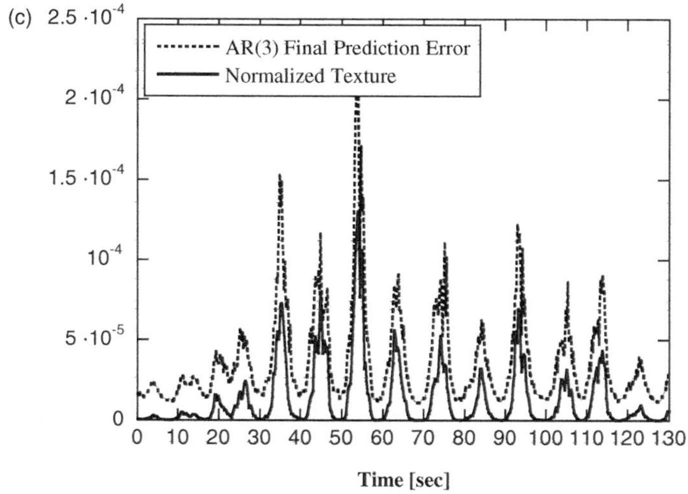

Figure 5.17 *(continued)*

$$M_{1I}(\alpha) = \lim_{N\to\infty} \frac{1}{N} \sum_{n=0}^{N-1} \mathbf{E}\left\{\tau(n)|x(n)|^2\right\} e^{-j\alpha n} = \lim_{N\to\infty} \frac{1}{N} \sum_{n=0}^{N-1} \mathbf{E}_\tau \left\{\tau(n) E_{x|\tau}\left\{|x(n)|^2\right\}\right\} e^{-j\alpha n} \tag{5.31}$$

Assuming that $\mathbf{E}_{x|\tau}\{|x(n)|^2\} = 1$ for all n, we may simplify (5.31) into

$$M_{1I}(\alpha) = \lim_{N\to\infty} \frac{1}{N} \sum_{n=0}^{N-1} \mathbf{E}\{\tau(n)\} e^{-j\alpha n} = M_{1\tau}(\alpha) \tag{5.32}$$

Therefore, $I(n)$ is first-order CS if, and only if, the texture $\tau(n)$ is itself first-order CS.[7] Since we do not have access to ensemble cyclic quantities, we estimate them from finite samples—that is, from $\{I(n)\}_{n=0}^{N-1}$ processing the intensity data. We estimate $M_{1I}(\alpha)$ from a single data record using the normalized fast Fourier transform (FFT) as follows:

$$\hat{M}_{1I}(\alpha) = \frac{1}{N} \sum_{n=0}^{N-1} I(n) e^{-j\alpha n} \tag{5.33}$$

In Fig. 5.18, the cyclic mean of the intensity data $Y(n)$, namely, $\hat{M}_{1Y}(\alpha)$, is plotted versus the cycle frequency α. It was estimated in the same way as in (5.10) by processing $N = 16{,}384$ samples of the first range cell. We observe that the most significant cycles are in the range $|\alpha| < 3.125 \times 10^{-3}$ (3.125 Hz), and the value in $\alpha = 0$ is zero. For a stationary process, we should find $\hat{M}_{1Y}(\alpha) \cong 0$ for every α, then the texture is indeed a cyclostationary process.

[7] The same result has been obtained in reference 16, assuming that texture and speckle are independent processes.

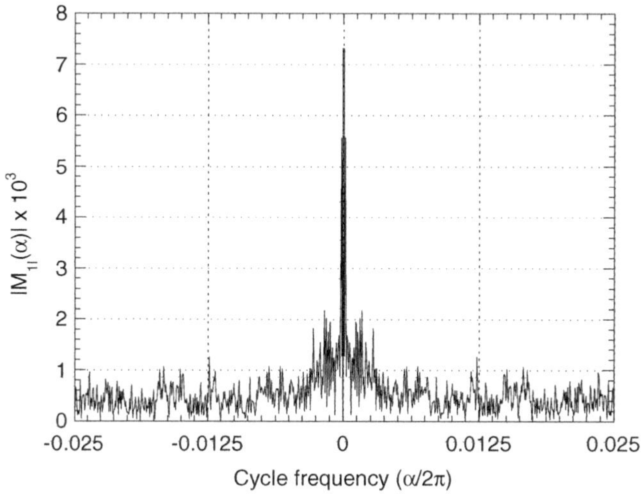

Figure 5.18 Sample estimate of the cycle mean of $Y(n)$ plotted in the range $\alpha \in [0.025, 0.025]$, sea clutter data from the first range cell of the file Starea12.

Due to the presence of waves with different wavelengths, we can model the non-negative sea-clutter texture as a sum of K cosinusoids and a constant term:

$$\tau(n) = A_0 + \sum_{k=1}^{K} A_k \cos(2\pi f_k n + \theta_k) \tag{5.34}$$

under the following assumptions:

(a1) A_k's are deterministic positive constants and A_0 is such that $\tau(n) \geq 0$ for every n;

(a2) θ_k's are deterministic constants in $[-\pi, \pi]$; and

(a3) f_k's are distinct and nonzero constants in $[-0.5, 0.5]$.

The intensity $I(n)$ can then be modeled as

$$I(n) = \tau(n) \cdot |x(n)|^2 = \left[A_0 + \sum_{k=1}^{K} A_k \cos(2\pi f_k n + \theta_k) \right] \cdot |x(n)|^2 \tag{5.35}$$

where it is assumed that

(a4) $x(n)$ is a complex Gaussian process with zero mean and unit conditional power.

The problem of interest herein is to retrieve $\tau(n)$ from $\{I(n)\}_{n=0}^{N-1}$; this amounts to estimating $\{A_k\}_{k=0}^{K}$, $\{f_k\}_{k=0}^{K}$, and $\{\theta_k\}_{k=0}^{K}$. First, A_0 is estimated as $\hat{A}_0 = \hat{M}_{1I}(0) = N^{-1}\Sigma_{i=0}^{N-1} I(i)$. To estimate the other parameters, we observe that under assumptions (a1)–(a4), the time-varying mean and the cycle mean of the intensity data $Y(n) = I(n) - \hat{A}_0$ are respectively given by

$$m_{1Y}(n) = \mathbf{E}\{Y(n)\} = \tau(n)\mathbf{E}_\tau\left\{\mathbf{E}_{x|\tau}\left\{|x(n)|^2\right\}\right\} - N^{-1}\sum_{i=0}^{N-1}\tau(i)\mathbf{E}_\tau\left\{\mathbf{E}_{x|\tau}\left\{|x(n)|^2\right\}\right\}$$
$$\approx \tau(n) - A_0 = \sum_{k=1}^{K} A_k \cos(2\pi f_k n + \theta_k) \tag{5.36}$$

and

$$m_{1Y}(n) = \lim_{n\to\infty}\frac{1}{N}\sum_{n=0}^{N-1} m_{1Y}(n)e^{-j\alpha n} = \sum_{k=1}^{K}\left(\frac{A_k}{2}e^{j\vartheta_k}\delta(\alpha - 2\pi f_k) + \frac{A_k}{2}e^{j\vartheta_k}\delta(\alpha + 2\pi f_k)\right) \tag{5.37}$$

where $\alpha \in (-\pi, \pi)$, and $\delta(\cdot)$ denotes the Kronecker delta function. The approximation $\mathbf{E}\{\hat{A}_0\} = N^{-1}\Sigma_{i=0}^{N-1}\tau(i) \approx A_0$ is accurate as long as $N \gg 1/f_k$ for each k. In fact, $\hat{A}_0$ is only asymptotically ($N \to \infty$) unbiased, whereas it is biased for a finite size of samples. Equation (5.37) reveals that the parameters of interest can be retrieved from the estimate $\hat{M}_{1Y}(\alpha)$ of the cycle mean as follows:

$$\{\hat{f}_k\}_{k=1}^{K} = \frac{1}{2\pi}\arg\max_{a_1, a_2, \ldots, a_k}\sum_{k=1}^{K}\left|\hat{M}_{1Y}(\alpha_k)\right|, \text{ with } \{\alpha_k\}_{k=1}^{K} \in (0, \theta)$$
$$\hat{A}_k = 2\left|\hat{M}_{1Y}(2\pi\hat{f}_k)\right|, \qquad \hat{\theta}_k = \angle\hat{M}_{1Y}(2\pi\hat{f}_k) \tag{5.38}$$

According to (5.34), the cyclic mean is estimated from the data as follows:

$$\hat{M}_{1Y}(\alpha) = N^{-1}\Sigma_{n=0}^{N-1}Y(n)e^{-j\alpha n}.$$

To find the locations of the K highest peaks in $\hat{M}_{1Y}(\alpha)$ in $(0, \pi]$, we use a relaxation-based optimization approach, referred to as the CM-RELAX algorithm [16]. The most important feature of the RELAX algorithm is that it decouples the multidimensional maximization problem in (5.38) into a series of one-dimensional maximization problems. The steps of the algorithm are summarized in reference 16.

We applied the CM-RELAX algorithm to the data $Y(n)$, obtained from the HH polarized sea-clutter data from the first range cell. Frequencies, amplitudes, and phases were estimated and used to retrieve the sea texture as described in (5.36). To verify the goodness of the method, we compared the model-based texture estimate to the nonparametric texture estimate obtained by the moving window estimator of (5.20). We retrieved the texture by using different values of N, K, and L. The results ploited in Fig. 5.19 have been obtained by processing N = 16,384 samples, and the other parameters used in the computation are K = 24 and L = 512. CM-RELAX locates all the frequency estimates in the range [−0.0125, 0.0125] (−12.5 Hz, 12.5 Hz). The results show a very good agreement between the two estimates. Note that while CM-RELAX estimates one for all the texture parameters, the MW estimator has to be implemented sequentially N times, each time L data have to be processed as described in (5.20). Moreover, the MW estimator does not allow reconstructing the texture outside the observation interval, whereas CM-RELAX does so.

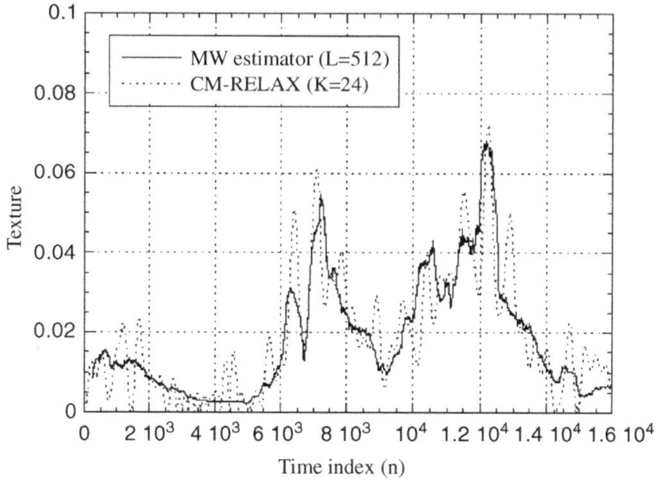

Figure 5.19 Texture retrieved by the CM-RELAX algorithm (5.39) and by the nonparametric MW estimator (5.20) for $N = 16{,}384$, $K = 24$, $L = 512$.

5.7 DISCUSSION

This chapter investigated the modulation induced on the sea Bragg scatterers by the long waves causing the long-term, nonstationarity of sea clutter. To this end, the numerical results reported herein concern data in down-wind conditions in a fully developed sea, where Bragg scattering is dominant. This dominance of the Bragg scattering was verified by estimation of mean power in each range cell for *HH* and *VV* polarized data and by computation of the polarization ratio. In accordance with the two-scale model, the *HH* power is always below the *VV* power and the polarization ratio is usually below unity. Inspired by the two-scale electromagnetic model and by the statistical compound-Gaussian model, we first performed a statistical analysis of the data amplitude. It was found that the data are very well fitted by the Generalized-*K* model, which is a particular case of the compound-Gaussian model. We then focused our analysis on *VV* data. For each range cell, we estimated the texture values, the time-varying spectrum centroid, and bandwidth. Moreover, to measure the mutual relationship between texture, centroid and bandwidth, we calculated the cross-covariance function between texture and centroid and between texture and bandwidth. Our results clearly show that texture, Doppler centroid, and bandwidth exhibit a common periodic behavior, with a period approximately equal to those of long sea waves, about 10 s. Based on our study, we conclude that, at least when Bragg scattering is dominant, long waves modulate the speckle scattering in amplitude, causing the variation of the local power; and in frequency, causing a periodic variation of the spectrum centroid and bandwidth.

5.7.1 Autoregressive Modeling of Sea Clutter

In an attempt to model the sea-clutter process, we also investigated the use of time-varying autoregressive (AR) modeling. We found a good compromise between complexity and accuracy with the AR(3) model, which is found to capture the physical phenomena being analyzed: (a) the periodicity of texture, centroid, and bandwidth and (b) the long-term, nonstationarity of sea clutter.

To be specific, we found that an AR model of order 3 provides a good compromise between model complexity and model accuracy in capturing the long-term nonstationary behavior of sea clutter. In so doing, we have further enriched the findings on AR modeling of sea clutter that were discussed in Chapter 4.

5.7.2 Cyclostationarity of Sea Clutter

After nonparametrically analyzing the behavior of the texture and its relationship with the Doppler centroid and bandwidth of the clutter spectrum, we performed a parametric analysis and modeling of the texture based on its cyclostationarity. We modeled the texture process as a multicomponent cosinusoidal signal and retrieved amplitudes, frequencies, and phases of the components by means of the CM-RELAX method. This estimator revealed non-negligible components with frequencies comprised in a range of few hertz around the frequency of the long waves. Then more insights into the nature of the sea backscatter were obtained by studying the relation between textures (estimated and parametrically reconstructed by the RELAX algorithm), Doppler centroid, and bandwidth. Cyclostationarity provides a distinct property of sea clutter that could be used in the radar scene characterization of a marine environment. Moreover, it may provide another discriminant for distinguishing between two inputs: sea clutter acting alone, and a target signal in the presence of clutter.

REFERENCES

1. S. Haykin, R. Bakker, and B. W. Currie (2002). Uncovering nonlinear dynamics—The case study of sea clutter, *Proc. IEEE* **90**(5), 860–881.
2. G. R. Valenzula (1978). Theories for the interaction of electromagnetic waves and oceanic waves—A review, *Bound. Layer Meteorol.* **13**(1–4), 61–65.
3. W. J. Plant and W. C. Keller (1990). Evidence of Bragg scattering in microwave Doppler spectra of sea return, *J. Geophys. Res.* **95**(C9), 16299–16310.
4. J. W. Wright (1968). A new model for sea clutter, *IEEE Trans. Antennas Propag.* **AP-16**, 217–223.
5. F. G. Bass, L. M. Fuks, A. E. Kalmykov, I. E. Ostrovsky, and A. D. Rosenberg (1968). Very high frequency radiowave scattering by a disturbed sea surface, *IEEE Trans. Antennas Propag.* **AP-16**, 554–568.
6. W. J. Plant and W. C. Keller (1983). The Two-Scale radar wave probe and SAR imagery of the ocean, *J. Geophys. Res.* **88**(C14), 9776–9784.

7. C. L. Rino, E. Eckert, A. Siegel, T. Webster, A. Ochadlick, M. Rankin, and J. Davis (1997). X-Band Low-Grazing-Angle Ocean Backscatter Obtained During LOGAN 1993, *IEEE J. Oceanic Eng.* **22**(1), 18–26.
8. A. D. Rozenberg, D. C. Quigley, and W. K. Melville (1996). Laboratory study of polarized microwave scattering by surface waves at grazing incidence: The influence of long-waves, *IEEE Trans. Geosci. Remote Sensing* **34**(6), 1331–1342.
9. D. Walker (2000). Experimentally motivated model for low grazing angle radar Doppler spectra of the sea surface, *IEE Proc. Radar Sonar Navig.* **147**(3), 114–120.
10. D. Walker (2001). Doppler modelling of radar sea clutter, *IEE Proc. Radar Sonar Navig.* **148**(2), 73–80.
11. T. J. Barnard and D. D. Weiner (1996). Non-Gaussian clutter modeling with generalized spherically invariant random vectors, *IEEE Trans. Signal Process.* **44**(10), 2384–2390.
12. E. Conte and M. Longo (1987). Characterisation of radar clutter as a spherically invariant random process, *IEE Proc.—F* **134**(2), 191–197.
13. A. Farina, F. Gini, M. Greco, and L. Verrazzani (1997). High resolution sea clutter data: A statistical analysis of recorded live data, *IEE Proc.—F* **144**(3), 121–130.
14. F. Gini and A. Farina (2002). Vector subspace detection in compound-gaussian clutter. Part I: survey and new results, *IEEE Trans. Aerosp. Electron. Syst.* **38**(4), 1295–1311.
15. T. Nohara and S. Haykin (1991). Canadian East Coast radar trials and the *K*-distribution, *IEE Proc.—F* **F138**(2), 80–88.
16. F. Gini and M. Greco (2002). Texture modelling, estimation, and validation using measured sea clutter data, *IEE Proc. Radar Sonar Navig.* **149**(3), 115–124.
17. E. Conte, M. Lops, and G. Ricci (1995). Asymptotically optimum radar detection in compound-Gaussian clutter, *IEEE Trans. Aerosp. Electron. Syst.* **31**, 617–625.
18. F. Gini, M. Greco, and L. Verrazzani (1995). Detection problem in mixed clutter environment as a Gaussian problem by adaptive pre-processing, *Electron. Lett.* **31**(14), 1189–1190.
19. F. Gini and M. Greco (1999). A suboptimum approach to adaptive coherent radar detection in compound-Gaussian clutter, *IEEE Trans. Aerosp. Electron. Syst.* **35**(3), 1095–1104.
20. K. J. Sangston, F. Gini, M. V. Greco, and A. Farina (1999). Structures for optimum and suboptimal coherent radar detection in compound-Gaussian clutter, *IEEE Trans. Aerosp. Electron. Syst.* **35**(2), 445–458.
21. D. B. Trizna (1985). A model for Doppler peak spectral shift for low grazing angle sea scatter, *IEEE J. Oceanic Eng.* **10**(4), 368–375.
22. P. H. Y. Lee, J. D. Barter, K. L. Beach, C. L. Hindman, B. M. Lake, H. Rungaldier, J. C. Shelton, A. B. Williams, R. Yee, and H. C. Yuen (1995). X-band microwave backscattering from ocean waves, *J. Geophys. Res.* **100**(C2), 2591–2611.
23. F. A. Fay, J. Clarke, and R. S. Peters (1977). Weibull distribution applied to sea-clutter, *Proceedings of the IEE Conference on Radar '77*, London, pp. 101–103.
24. E. Jakeman and P. N. Pusey (1976). A model for non-Rayleigh sea echo, *IEEE Trans. Antennas Propag.* **AP-24**, 806–814.
25. K. D. Ward, C. J. Baker, and S. Watts (1990). Maritime surveillance radar. Part I: Radar scattering from ocean surface, *IEE Proc.—F* **137**(2), 51–62.
26. D. M. Drumheller and H. Lew (2002). Homodyned-*K* fluctuation model, *IEEE Trans. Aerosp. Electron. Syst.* **38**(2), 527–542.
27. H. Lew and D. M. Drumheller (2002). Estimation of non-Rayleigh clutter and fluctuating-target models, *IEE Proc., Radar, Sonar Navig.* **149**(5), 231–241.
28. H. D. Griffiths (2003). Knowledge-based solutions as they apply to the general radar problem, in *Proceedings of the RTO NATO Lecture Series 233 on Knowledge-Based Radar Signal and Data Processing*, Rome, Italy, 6–7 November, 2003.
29. M. Greco, F. Bordoni, F. Gini, and L. Verrazzani (2003). *X-band sea* clutter non-stationarity: The *influence of long waves*, Technical Report, University of Pisa, Pisa, Italy, July 2003.
30. S. Haykin (1991). *Advances in Spectrum Analysis and Array Processing.* Vol. I, Simon Haykin (ed.), Prentice-Hall, Englewood Cliffs, NJ.

31. C. W. THERRIEN (1992). *Discrete Random Signals And Statistical Signal Processing*, Prentice-Hall, Englewood Cliffs, NJ.
32. S. HAYKIN AND D. J. THOMSON (1998). Signal detection in a nonstationary environment reformulated as an adaptive pattern classification problem, *IEEE Proc.* **86**(11), 2325–2344.
33. S. SHAMSUNDER, G. B. GIANNAKIS, AND B. FRIEDLANDER (1995). Estimating random amplitude polynomial phase signals: A cyclostationary approach, *IEEE Trans. Signal Process.* **43**(2), 492–505.
34. G. ZHOU AND G. B. GIANNAKIS (1995). Harmonics in multiplicative and additive noise: Performance analysis of cyclic estimators, *IEEE Trans. Signal Process.* **43**(6), 1445–1460.
35. A. DROSOPOULOS (1994). Description of the ohgr database, Technical note No. 94-14, Defence Research Establishment Ottawa, December 1994.

Chapter 6

Two New Strategies for Target Detection in Sea Clutter†

Rembrandt Bakker, Brian Currie, and Simon Haykin

6.1 INTRODUCTION

The tracking of targets in a radar search space is basic to any radar surveillance application. The search space is typically spanned by a subset of six dimensions: time, frequency, Doppler shift, range, azimuth, and elevation. By virtue of the discrete sampling of all measured dimensions, the search space is partitioned into a set of *resolution cells*. The measurement of all resolution cells at a given discrete time t constitutes a *frame*. A wide variety of algorithms have been designed to perform target tracking. Typically, these algorithms involve three major steps:

1. Collect a frame of data.
2. For each resolution cell, decide whether or not the frame contains a target.
3. Use the detections to initiate, continue or terminate target tracks.

The three best-known algorithms to perform step 3 are the Nearest-Neighbor (NN) algorithm [1], Probabilistic Data Association (PDA) [2], and Multiple

† The material presented in this chapter is based on the following:

1. R. Bakker, B. Currie, T. Kiribarajan, and S. Haykin, (2002). Adaptive radar detection: A Bayesian approach, EPSRC and IEE Workshop on Nonlinear and Non-Gaussian Signal Processing, Peebles, Scotland, July 8–12, 2002.
2. R. Bakker, G. Lopez-Risueno, and S. Haykin, Bayesian approach to the direct filtering of radar targets in clutter (Internal Report), Adaptive Systems Laboratory, McMaster University, August 2002.
3. B. Currie and S. Haykin (2004) ((DRDC Report), Bayesian Detector Evaluation and Comparison, Final Report prepared for Defence Research and Development, Ottawa, under contract No. W7714-020683/001/SV, pp. 1–58, April 2004. Permission of the DRDC to reproduce results from this report is deeply appreciated.

Adaptive Radar Signal Processing. Edited by Simon Haykin

Hypothesis Tracking (*MHT*) [3]. The NN (not to be confused with the abbreviation NN used for a neural network in Chapter 3) algorithm updates the tracking filter with the measurement closest to the predicted state. In PDA, each measurement affects the tracking filter to a degree based on the probability that it is correct given the predicted state. MHT creates a new hypothesis track for every measurement that is within a validation region, and then it prunes unlikely tracks using a likelihood ratio test.

Each of these algorithms uses double thresholds, first for performing the detections in step 2 and then for the acceptance/rejection of tracks in step 3. Speaking in computational terms, the intermediate detections are attractive, because typically only a small fraction of all measurements will be developed as detection. But from an information-theoretic point of view, the intermediate detection step (based on hard decisions) wastes important information; we say so because a real number lying between 0 and 1 is rounded off to an integer. This problem has been recognized in the literature, and as a first attempt to alleviate it, the simple (NN) algorithm was modified into the Strongest Neighbor (SN) algorithm [4]. This latter algorithm does intermediate detections, but retains the amplitudes of the original measurements. It augments tracks by selecting, among neighbors that are reasonably close, the one with the strongest amplitude. The strong gains reported for the SN algorithm make us pose the following question: Is it computationally feasible to use an algorithm which avoids intermediate detections all together and works in the measurement space directly?

Henceforth, we will refer to such an algorithm as a *direct tracking algorithm*. The direct tracking algorithm consists of three steps of its own:

(i) Collect radar returns from a given search area for a certain period of time.

(ii) For each resolution cell, recursively compute the probability that it contains a target.

(iii) Detect target tracks from the evolution of the target probability distribution with time.

When only the first two steps are considered, we speak of a *direct filtering algorithm*.

A Bayesian approach to the direct tracking problem was developed by Bruno and Moura [5]. Their algorithm tracks M possible targets with known signatures in a search space of R resolution cells. The algorithm first computes the probability of each of the 2^M different target combinations. The centroid of each target can be in any of the R resolution cells or can be absent. Bruno and Moura claim a huge improvement, up to 11 dB, over a tracker that uses intermediate detections and Kalman filters. For a single target, the computational complexity of the prediction step is $O(R^2)$; if the mobility of the target is restricted to nearby resolution cells only, then the computational complexity is reduced to $O(R)$.

In the first part of this chapter, following references 7 and 8, we present a different Bayesian direct filtering algorithm, which applies Bayes' rule in such a way that it can also be used as a *smoother*. In the smoothing mode, the filter output is conditioned on both past and future measurements. We apply it to data from our large collection of real-world sea-clutter data, which is known as the Dartmouth database and which was collected with the IPIX radar. The results are very encouraging: The filter shows the target even for a dataset with an average target-to-clutter ratio well below 0 dB. For a single-target scenario, the filter can run in real time. The handling of a search space with many dimensions is straightforward.

In the second part of this chapter, we present another detection algorithm, rooted in stochastic differential equation (SDE) theory. The algorithm, known as the *correlation anomaly* algorithm, is due to Field [9, 10]. The rationale of this second algorithm is based on predictive modeling, where the SDE theory is used to predict the evolution of sea clutter. Any anomalous deviation of the actual received signal from the predictive model is attributed to the possible presence of a target.

The chapter finishes with a comparative evaluation of these two new detection strategies using real-life radar data.

6.2 BAYESIAN DIRECT FILTERING PROCEDURE

Given current and past measurements, the task of the Bayesian direct filtering procedure is to compute, for each resolution cell, the *probability that the cell under test contains a target.* In this section, we focus the discussion on a single-target scenario.

6.2.1 Single-Target Scenario

Let it be assumed that the multi-dimensional measurement space is indexed into a linear array of resolution cells. The output of the Bayesian filter is a vector whose elements contain the probability $P\{\in_t^r|\mathbf{z}_t, \mathbf{Z}_t^{\text{past}}\}$, where $\in_t^r$ is the event that a target resides in resolution cell r at time t, $\mathbf{z}_t$ is the current frame, and $\mathbf{Z}_t^{\text{past}}$ is the set of all previous frames. A special case is $\in_t^0$, which refers to the event that there is no target. The application of *Bayes' rule* yields the *a posteriori* probability

$$P\left\{\in_t^r \middle| \mathbf{z}_t, \mathbf{Z}_t^{\text{past}}\right\} = \frac{p\left(\mathbf{z}_t, \mathbf{Z}_t^{\text{past}} \middle| \in_t^r\right) P\left\{\in_t^r\right\}}{p\left(\mathbf{z}_t, \mathbf{Z}_t^{\text{past}}\right)} \tag{6.1}$$

where in the numerator, $p(\mathbf{z}_t, \mathbf{Z}_t^{\text{past}}|\in_t^r)$ is the conditional probability density function (pdf) of the joint-data-pair $(\mathbf{z}_t, \mathbf{Z}_t^{\text{past}})$, given the event $\in_t^r$, and $P(\in_t^r)$ is the *prior probability* of the event $\in_t^r$; in the denominator, $p(\mathbf{z}_t, \mathbf{Z}_t^{\text{past}})$ is the pdf of the joint data-pair $(\mathbf{z}_t, \mathbf{Z}_t^{\text{past}})$. The denominator, defined by

$$p\left(\mathbf{z}_t, \mathbf{Z}_t^{\text{past}}\right) = \sum_r p\left(\mathbf{z}_t, \mathbf{Z}_t^{\text{past}} \middle| \in_t^r\right) P\left(\in_t^r\right) \tag{6.2}$$

serves the role of a normalizing factor.

We assume that the past and current measurements depend only on the presence of the target and not on each other, so that we may expand (6.1) to write

$$P\left\{\in_t^r \middle| \mathbf{z}_t, \mathbf{Z}_t^{\text{past}}\right\} = \frac{p\left(\mathbf{z}_t \middle| \in_t^r\right) p\left(\mathbf{Z}_t^{\text{past}} \middle| \in_t^r\right) P\left\{\in_t^r\right\}}{p\left(\mathbf{z}_t, \mathbf{Z}_t^{\text{past}}\right)} \tag{6.3}$$

The new term $p(\mathbf{z}_t|\in_t^r)$ in the numerator of (6.3) requires models that specify the statistical behavior of the measurement for the clutter-only and clutter-plus-target scenarios. These models are presented in Sections 6.3.2 and 6.3.3 for the special case of the detection of small floating objects in sea clutter. If several measurements of the same variable are available, the term $p(\mathbf{z}_t|\in_t^r)$ expands in a corresponding way. For example, if we have two measurements $\mathbf{z}_t^{(1)}$ and $\mathbf{z}_t^{(2)}$, this term becomes $p(\mathbf{z}_t^{(1)}, \mathbf{z}_t^{(2)}|\in_t^r)$. If the measurements only depend on the hidden state $\in_t^r$ and not directly on each other, then this term simplifies to $p(\mathbf{z}_t^{(1)}|\in_t^r)p(\mathbf{z}_t^{(2)}|\in_t^r)$.

The conditional probability density function $p(\mathbf{Z}_t^{\text{past}}|\in_t^r)$ in the numerator of (6.3) is calculated using the recursive formula:

$$p\left(\mathbf{Z}_t^{\text{past}} \middle| \in_t^r\right) = \sum_{q=0}^{R} p\left(\mathbf{Z}_t^{\text{past}} \middle| \in_{t-1}^q, \in_t^r\right) P\left(\in_{t-1}^q \middle| \in_t^r\right) \tag{6.4a}$$

$$= \sum_{q=0}^{R} p\left(\mathbf{z}_{t-1}, \mathbf{Z}_{t-1}^{\text{past}} \middle| \in_{t-1}^q\right) P\left(\in_{t-1}^q \middle| \in_t^r\right) \tag{6.4b}$$

$$= \sum_{q=0}^{R} p\left(\mathbf{z}_{t-1} \middle| \in_{t-1}^q\right) p\left(\mathbf{Z}_{t-1}^{\text{past}} \middle| \in_{t-1}^q\right) P\left(\in_{t-1}^q \middle| \in_t^r\right) \tag{6.4c}$$

Equation (6.4a) introduces the *transition matrix* $P(\in_{t-1}^q|\in_t^r)$. It contains the probabilities that event q preceded event r. The specification of these probabilities requires a target model that typically contains information on the agility of the target: The less agile the target, the smaller the jumps it can be expected to make in resolution space and the sparser the transition matrix will therefore be; see Section 6.3.4 for an example.

6.2.2 Conditioning on Past and Future Measurements

The filtering equations (6.3) and (6.4) differ from the usual predict/update equations in sequential state estimation. Specifically, applying Bayes' rule to the probability $P(\in_t^r|\mathbf{z}_t, \mathbf{Z}_t^{\text{past}})$ on the left-hand side of (6.3) yields

$$P\left\{\in_t^r \middle| \mathbf{z}_t, \mathbf{Z}_t^{\text{past}}\right\} = \frac{p\left(\mathbf{z}_t \middle| \in_t^r, \mathbf{Z}_t^{\text{past}}\right) P\left\{\in_t^r \middle| \mathbf{Z}_t^{\text{past}}\right\}}{p\left(\mathbf{z}_t \middle| \mathbf{Z}_t^{\text{past}}\right)} \tag{6.5a}$$

$$= \frac{p\left(\mathbf{z}_t \middle| \in_t^r\right) P\left\{\in_t^r \middle| \mathbf{Z}_t^{\text{past}}\right\}}{p\left(\mathbf{z}_t \middle| \mathbf{Z}_t^{\text{past}}\right)} \tag{6.5b}$$

In this new formulation, we first note that a prediction is made to obtain $P\{\in_t^r|\mathbf{Z}_t^{\text{past}}\}$. The prediction is then updated with the current measurement to form the filtered output $P\{\in_t^r|\mathbf{z}_t, \mathbf{Z}_t^{\text{past}}\}$. The prediction posterior probabilities can be obtained by a recursive equation delineated in reference 5, which uses the same transition matrix as (6.4a). The reason that we opt for the alternative filtering scheme is that it is easier to expand it to the case where the filtering is conditioned on both past and future measurements. The filter is then called a *smoother*. In the smoothing mode, (6.3) becomes

$$p\{\in_t^r | \mathbf{z}_t, \mathbf{Z}_t^{\text{past}}, \mathbf{Z}_t^{\text{future}}\} = \frac{p(\mathbf{z}_t | \in_t^r) p(\mathbf{Z}_t^{\text{past}} | \in_t^r) p(\mathbf{Z}_t^{\text{future}} | \in_t^r) P\{\in_t^r\}}{p(\mathbf{z}_t, \mathbf{Z}_t^{\text{past}}, \mathbf{Z}_t^{\text{future}})} \tag{6.6}$$

The extra factor $p(\mathbf{Z}_t^{\text{future}} | \in_t^r)$ in the numerator of (6.6) can be computed by running the recursive equation (6.4c) backwards in time.

6.3 OPERATIONAL DETAILS

6.3.1 Experimental Data

We make use of an extensive set of experimental data collected with the McMaster IPIX radar at the east coast of Canada, in Dartmouth, Nova Scotia. For the subset of the database that we use, the radar is operated in the dwell mode with a 1° pencil beam and a fixed radio frequency of 9.39 GHz. It is mounted on a cliff top at 30 m above sea level. The object observed by the radar is a 1 m-diameter *styrofoam ball* which is wrapped in radar reflective material and which floats on the ocean surface on an anchor line at about 2.5 km off shore. Range is sampled at 15-m intervals, but the actual range resolution is 30 m, obtained by using a 200-ns rectangular pulse. The actual pulse repetition frequency (PRF) is 2000 Hz, but the pulse alternates between horizontal (*H*) and vertical (*V*) polarization, so that the effective (and documented) PRF is 1000 Hz. For each pulse, both *H* and *V* polarizations are recorded simultaneously, resulting in four possible transmit/receive polarization combinations: *HH*, *HV*, *VH*, and *VV*. For each combination, the amplitude and phase of the radar returns are stored as in-phase (*I*) and quadrature (*Q*) components.

6.3.2 Statistics of Sea Clutter

In this chapter, we apply the Bayesian direct filter to the floating object in the Dartmouth database as seen in the time-Doppler image. Figures 6.3a–6.5a show typical examples. Note that Doppler frequency and Doppler velocity are linearly related. At any instant of time, the floating object has a fixed velocity, appearing as a sharp line in the image. The waves cause the target to alternate between moving away from and toward the radar. It has zero average Doppler velocity because it is anchored. The underlying sea clutter has a much wider velocity distribution. Its average Doppler velocity depends on the wind direction with respect to the radar beam.

We distinguish two ways to discriminate target from clutter in the Doppler spectrum: (1) based on the difference in signal strength, and (2) based on the difference in signal shape. In our case, the strength of the target signal is comparable to that of the clutter signal. A detector based on signal strength would therefore only be possible if the statistics of the clutter's power spectrum are well known. Unfortunately, the statistical distribution of the power in each bin of the Doppler spectrum is very hard to estimate. In the short term up to 1000 ms, it can be adequately modeled by a low-order complex AR process [6], but on the longer term this process is nonstationary and it is also highly dependent on the sea state. In our case, with a weak target, it is better to do the detection based on the difference in signal shape. Sea clutter always has a smooth spectrum, whereas the small target appears as a line component. This concept is illustrated in Fig. 6.1a. To highlight the line component, we transform the power spectrum by dividing each value in it by the mean value of its b nearest neighbors. This will accentuate the narrow peak of the line component and suppress the wide peak of the clutter (see Fig. 6.1b). We call the transform a "peak filter." It is inspired by the grouped periodogram test described by Priestley [11]. The statistics of the peak filter output can be assessed by making three assumptions:

1. None of the b neighbors contains a target.
2. The clutter is white within the window of length $b + 1$.
3. All $b + 1$ power spectrum ordinates are sampled independently.

If these three conditions are met, the normalized individual power spectrum coefficients have a χ^2 *distribution with two degrees of freedom* [11]. The peak filter output, which divides one coefficient by b others, then has an *F-distribution with (2, 2b) degrees of freedom.* Figure 6.2 shows the $F_{2,2b}$ distribution for the case of $b = 6$.

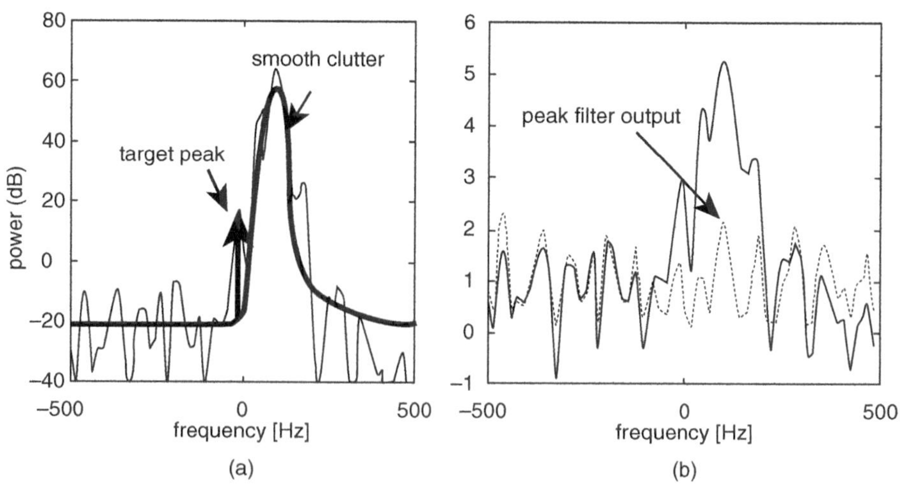

Figure 6.1 A 64-sample-based Doppler spectrum of target plus clutter signal. (a) Assuming the underlying clutter is smooth and the weak target has a line spectrum. (b) Peak filter output.

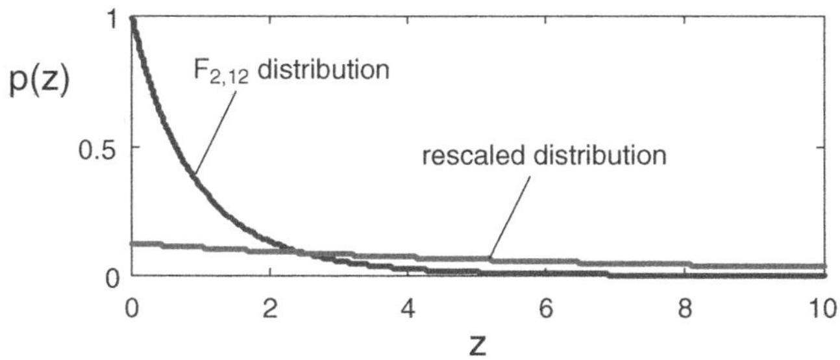

Figure 6.2 *F*-distribution with 2 and 12 degrees of freedom, along with the rescaled version.

Assumptions 1–3 are not always satisfied in practice. First, if any of the b neighbors contains a target, the peak filter output will be lowered. The peak filter therefore has difficulty to detect multiple targets if they are less then b Doppler bins apart. Second, the clutter power can vary quite abruptly with frequency, and to assume that it is locally white requires that the power spectrum be densely sampled. We find that for $b = 6$, the power spectrum needs to be obtained from at least 64 to 128 observations for *VV/HH* polarized clutter, respectively. Third, the independent sampling condition (i.e., condition 3) is not satisfied when a data taper other than the rectangular window is used. Nevertheless, we may employ a Hamming window to reduce the bias of the estimated spectrum.

Finally, we note that the peak filter compares power spectrum coefficients in relation to their neighbors in the Doppler spectrum.[1]

6.3.3 Statistics of Target Returns

When we are dealing with an unknown target, modeling of the target's statistics is not very straightforward. The best that we can do is to model the smallest target that we expect to be able to see. If we pick too weak a target, the Bayesian filter may easily confuse the target with clutter. If we pick too strong a target, the filter will only pick up strong targets. We also need to decide whether the target cross-section is constant or whether it is random. Given that the small floating objects often hide behind ocean waves, we opt for a random cross-section. For ease of implementation and lack of detailed knowledge about the target, we use the same distribution that also governs the sea clutter, but we rescale it to reflect the difference between the radar cross-section of clutter and that of the target. Specifically, if the clutter distribution is given by $F_{2,2b}(z)$, the target distribution is given by $\frac{1}{\gamma}F_{2,2b}(\frac{z}{b})$, where γ denotes target-to-clutter power ratio. Figure 6.2 shows the rescaled F-distribution for the case of $\gamma = 2\,\text{dB}$. In our experiments, we use $\gamma = 1\,\text{dB}$.

[1] *Two ways for future improvement are, first, the development of a statistical model for the raw clutter* spectrum and, second, the comparison of power spectrum coefficients across range.

6.3.4 Motion Model of the Target

The elements of the transition matrix $P\{\in_{t-1}^{q} \mid \in_{t}^{r}\}$ are determined by (1) a model for the agility of the target, and (2) *a priori* information regarding the presence and persistence of the target. We model the target as having a Gaussian distributed random acceleration, with standard deviation $\sigma = 2$ meters per second (m/s). The *a priori* information is incorporated as follows. When we look at a random patch of the ocean surface, we will sometimes see clutter, and sometimes see a target. We assume that the durations of seeing clutter and target both have an exponential distribution, each with its own median life time. We use a median life time of 100s for clutter, and also 100s for the target. This way, the *a priori* probability of seeing a target is 0.5. The transition probability $P\{\in_{t-1}^{q} \mid \in_{t}^{r}\}$ is now obtained by looking at the Doppler velocity difference Δv between the resolution cells q and r. The velocity difference Δv is rescaled into the units m/s^2, and the random acceleration distribution is applied. The result is then corrected for the probability that the target disappears, and the probability that a new target appears.

6.4 EXPERIMENTAL RESULTS ON THE BAYESIAN DIRECT FILTER

To undertake the experimental ability of the Bayesion filter, we chose three datasets from the Dartmouth database to test the Bayesian direct filter. For each dataset, the target-to-clutter ratio (TCR) is estimated using the formula

$$\text{TCR} = \frac{P_{\text{total}} - P_{\text{clutter}}}{P_{\text{clutter}}}$$

The clutter power P_{clutter} is obtained by averaging the signal power over those range bins that contain clutter only. The datasets are:

1. File \#30, 19931109191449starea.cdf, range bin 6, wind perpendicular to radar beam, TCR = −3 dB
2. File \#280, 19931118023604stareC0000, range bin 7, wind blowing toward radar, TCR = −7 dB
3. File \#311, 19931118162658stareC0000.cdf, wind blowing away from radar, TCR = 3 dB

Initially, we used only the *VV* polarization. Doppler spectra are estimated on a 64-sample sliding window basis. We distinguish three modes of the algorithm: *Single-frame mode* uses only the current frame to compute the probability $P\{\in_{t}^{r} | \mathbf{z}_t\}$; *forward filtering mode* is the actual Bayesian direct filter in which feedback is used to compute $P\{\in_{t}^{r} | \mathbf{z}_t, \mathbf{Z}_t^{\text{past}}\}$; *forward/backward mode* is the Bayesian direct smoother in which conditioning is done on both future and past measurements, as in $P\{\in_{t}^{r} | \mathbf{z}_t, \mathbf{Z}_t^{\text{past}}, \mathbf{Z}_t^{\text{future}}\}$. This shows the performance of the peak-filter only, without using any information from the target motion model. Figures 6.3a–6.5a show the raw time-Doppler spectra for the three datasets, and Figs. 6.3b–6.5b show the output of

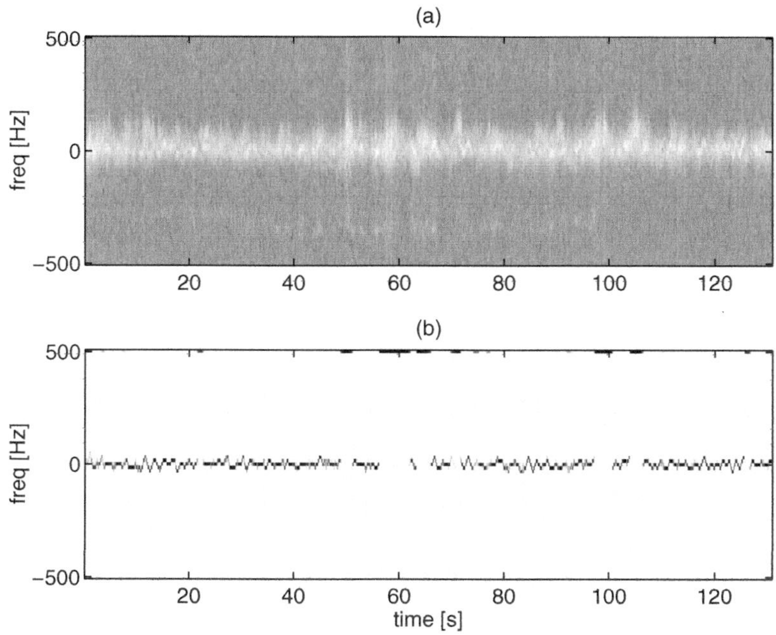

Figure 6.3 (a) A 64-sample sliding window time-Doppler image of dataset 1. (b) Result of Bayesian direct smoother. Each pixel represents the probability of a target in the corresponding resolution cell. The darker the pixel, the higher the probability of target detection. The top horizontal line of the image represents the probability that there is no target.

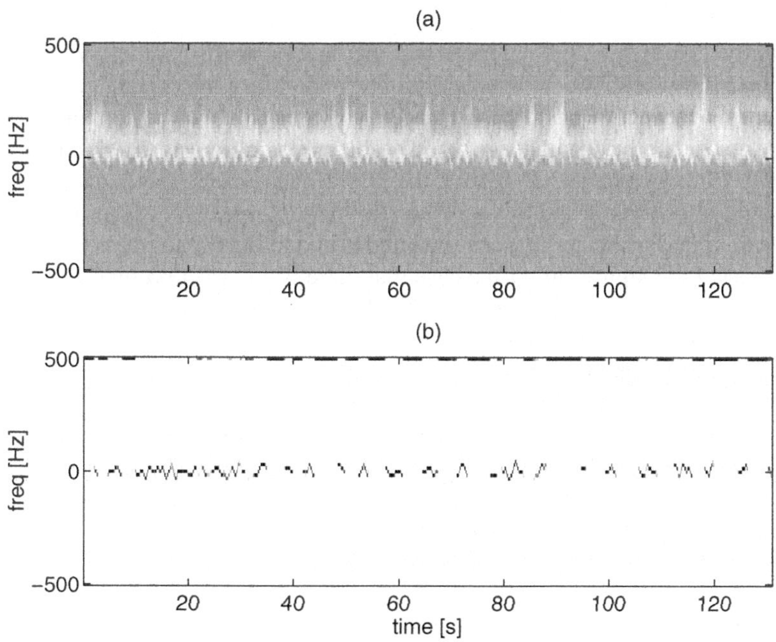

Figure 6.4 (a) Sample sliding window time-Doppler image of dataset 2. (b) Result of Bayesian direct smoother. Each pixel represents the probability of a target in the corresponding resolution cell. The darker the pixel, the higher the probability of target detection. The top horizontal line of the image represents the probability that there is no target.

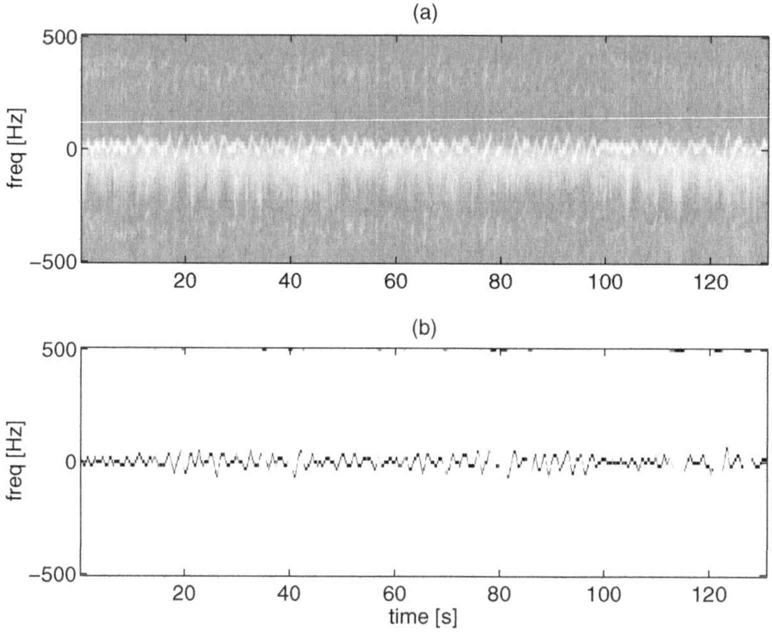

Figure 6.5 (a) A 64-sample sliding window time-Doppler image of dataset 3. (b) Result of Bayesian direct smoother. Each pixel represents the probability of a target in the corresponding resolution cell. The darker the pixel, the higher the probability of target detection. The top horizontal line of the image represents the probability that there is no target.

the Bayesian direct smoother. It is observed that the Bayesian direct smoother highlights the target really well. Whenever the filter does not detect the target, it can be seen from the time-Doppler image that the target is also invisible to the human eye.

We investigated the performance further by focusing on dataset 2, which has the lowest TCR. Figures 6.6a and 6.4b show the performance of the Bayesian direct filter operating in single-frame and forward-filtering mode, respectively. Not surprisingly, there is a large difference between the single- and multi-frame mode.

Next, we tested the algorithm for the case that there is no target present. Figure 6.7 shows the equivalent of Fig. 6.4 for range bin 4 of the same data file. This range bin is known not to contain any target. Only at time $t = 96\,\text{s}$, the filter assigns a small probability to the target.

As discussed in Section 6.2.1, the filter can be easily enhanced by including multiple measurements. Figure 6.8 shows the effect of including the *HH* measurements into the filter design. It shows the presence of the target for a much larger percentage of the time, but it also shows false tracks at around 45 s and 75 s. This is because the power spectrum of the *HH* polarization is narrower than that of *VV*, violating the assumptions underlying the *F*-distribution as discussed in Section 6.3.2.

To measure the performance of the Bayesian direct filter, we need to convert the filter output into a target track and compare it to the true target trajectory. To

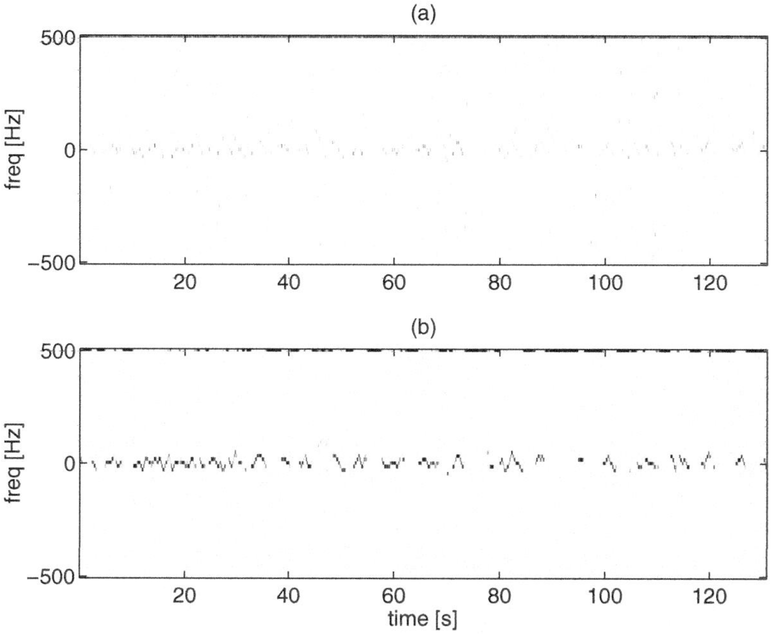

Figure 6.6 (a) Result of Bayesian direct filter operating in single-frame mode for dataset 2. (b) Result of Bayesian direct filter operating in forward-filtering mode. Compare to Fig. 6.4.

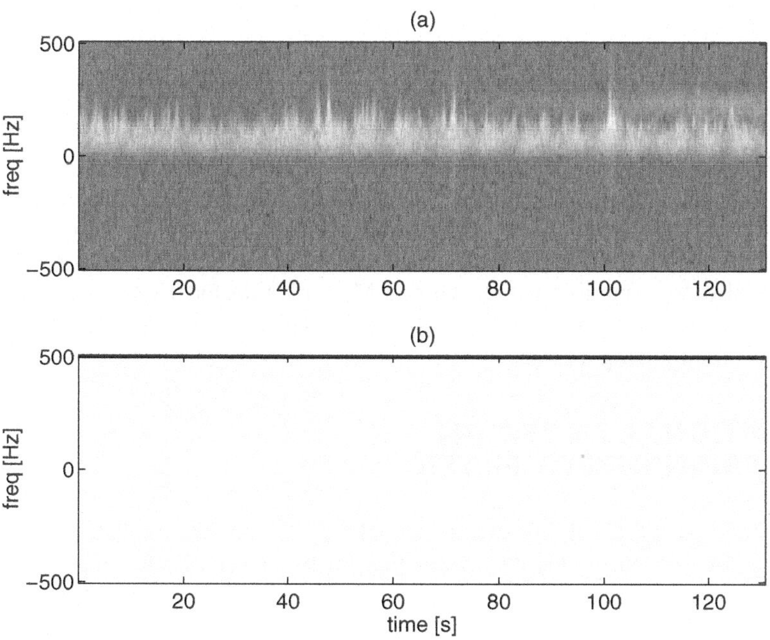

Figure 6.7 (a) A 64-sample sliding window time-Doppler image of file #280, range bin 4. (b) Result of Bayesian direct smoother. Compare to Fig. 6.4.

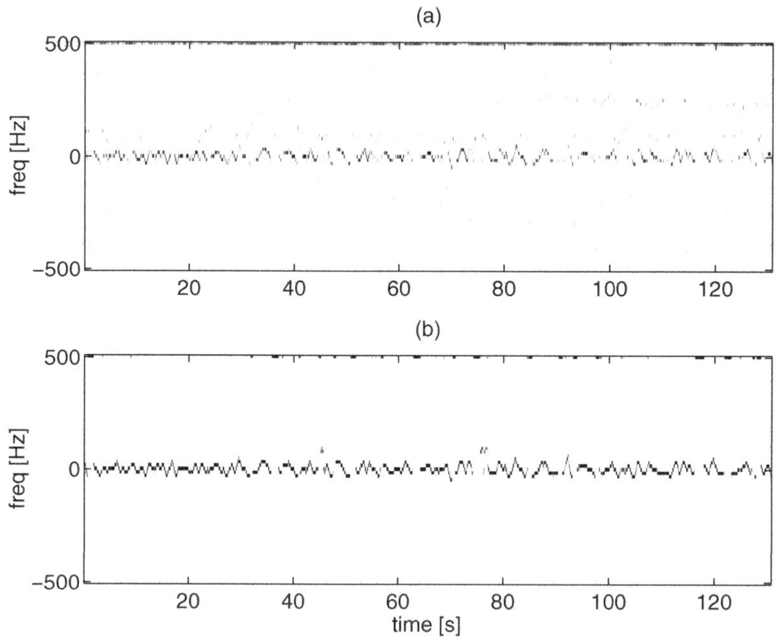

Figure 6.8 Effect of including the *HH* channel besides *VV* for dataset 2. (a) Result of Bayesian direct filter operating in single-frame mode. (b) Result of Bayesian direct filter operating in forward filtering mode. Compare to Fig. 6.6.

find the true trajectory of the object, we make use of the fact that although in file #280 the target appears in range bin 7, it actually shows up much stronger in range bin 8, at a TCR of 0 dB. The fact that it appears in multiple range bins is due to range oversampling and multipath effects. The signal in range bin 8 is strong enough to extract a target trajectory which, upon visual inspection, looks error-free. To convert the filter output into target tracks, we threshold it at a probability of 0.5. This is essentially the same as what Bruno and Moura [5] do for the single target case: They declare a detection when the likelihood of having a target is larger than the likelihood of not having a target.

6.5 ADDITIONAL NOTES ON THE BAYESIAN DIRECT FILTER

In this first study on the feasibility of target tracking without intermediate detections, we have focused on producing images that lend themselves for comparison with the human eye. We conjecture that a human operator could not do a much better job of following the target than the Bayesian direct filter, especially when running in the smoothing mode. This conjecture may have practical implications, since the Bayesian direct filter is easily extended to higher-dimensional problems whereas the human eye is not. But we need to keep in mind that the human operator acts as a fully adaptive system.

Current implementation of the direct filter depends on a number of operational details, the effect of which are highlighted here:

1. **Clutter Statistics**: The peak filter is a good work-around tool for the open problem of modeling the Doppler spectrum statistics. But if the clutter spectrum narrows, or the number of samples available per frame drops, the assumptions outlined in Section 6.3.2 may be violated.
2. **Target Statistics**: In Section 6.3.3 we set the pdf for the target signal strength equal to that of the clutter, except for a scaling factor which reflects that, on average, the target gives a stronger return than the clutter. This is a very general assumption that works for almost any type of target. We have run experiments in which we allow ourselves to tailor this pdf to the target at hand, by choosing a pdf with more density at the expected TCR. This greatly enhanced the detection performance, but we have not pursued this further because we do not want to fine-tune the filter to any particular data-set.
3. **Target Mobility**: The filter output is sensitive to the choice of the σ parameter, which determines the agility of the target. For low σ, the target is only seen when it is not accelerating. For high σ, on the other hand, the algorithm may easily confuse sequences of small clutter peaks with the target.
4. ***A priori* Probabilities**: If we set the *a priori* probability of having a target in one of the Doppler bins at p_0, any positive evidence toward the presence of a target will raise the *a posteriori* probability above p_0. More interesting are the target and clutter median lifetime settings. The shorter these lifetimes are, the quicker the filter will pick up a target, with the inherent danger of picking up a target before accumulating sufficient evidence.

6.6 CORRELATION ANOMALLY DETECTION STRATEGY

At the same Peebles workshop where the Bayesian direct filter [7] was first described, Field [9] also described a model for electromagnetic backscatter from random scatterers, which are representative of sea clutter. The backscattering model is formulated using *stochastic differential equation* (SDE) theory for the scattering cross-section, scattering intensity, and phase. Subsequently, a detailed description of the SDE-based model was presented in reference 10, where the idea of a new radar receiver, called the *correlation anomaly receiver*, was also described. The rationale for this new receiver is that it may be viewed as a form of *predictive modeling*, where the SDE theory is first used to predict the behavior of sea clutter acting alone; then, any anomalous deviation from that prediction is attributed to the possible presence of a target.

With the Bayesian direct filter and correlation anomaly receiver providing the frameworks for two new radar receivers (each being novel in its own way), it is apropos that we compare their performances, which is done in Section 6.7. Before proceeding to do that, however, a brief description of the correlation anomaly receiver is presented in what follows.

In reference 10, Field and Tough use SDE theory to demonstrate that the amplitude fluctuations of the scatterers can be well-modeled by the *K*-distribution; the *K*-distribution was discussed in Chapters 4 and 5. Specifically, a fundamental equation is derived for the squared volatility of the amplitude process, namely,

$$|d\psi_t|^2 = \left(B_{x_t} + \frac{Az_t}{2x_t} \right) dt \tag{6.7}$$

where ψ_t is the complex observed signal (referred to as the amplitude process in reference 10), x_t is the cross-section at time t, z_t is the intensity also at time t, and A and B are constants measured in units of frequency. The constant A corresponds to the time frame of variations in the cross-section x_t, while the constant B corresponds to the time-scale of the Rayleigh-distributed fluctuations; usually, we have $A << B$.

Field and Tough [10] suggest that over the short term, the cross-section x_t may be considered a constant, on the basis of which the signal exhibits only Rayleigh scattering. Hence, with all the terms on the right-hand side of (6.7) being constants except for z_t, it is observed that the left-hand and right-hand sides of the equation should be linearly correlated for a truly *K*-distributed process. Using the defining equation

$$c(X,Y) = \frac{\mathbf{E}(X - \mathbf{E}[X])(Y - \mathbf{E}[Y])}{(\mathrm{Var}[X] \cdot \mathrm{Var}[Y])^{1/2}} \tag{6.8}$$

as the correlation function, Field and Tough evaluate the new parameter

$$c_\psi = \left(|d\psi_t|^2, z_t \right) \tag{6.9}$$

which lies in the range [−1, 1]. By monitoring the value of parameter c_ψ, a framework is proposed for a *correlation anomaly receiver.* In particular, when c_ψ falls below some prescribed threshold, it indicates the occurrence of some anomaly in the received radar data, which is attributed to the presence of a target [10].

6.7 EXPERIMENTAL COMPARISON OF THE BAYESIAN DIRECT FILTER AND CORRELATION ANOMALY RECEIVER

For the experimental comparison of these two receivers, two sources of real-life coherent radar data were used in a detailed study reported in reference 12. The two datasets were

- McMaster IPIX radar data
- QinetiQ radar data

The IPIX radar data were collected at the Osborne Head Gunnery Range (OHGR) coastal site near Dartmouth, Nova Scotia; this database was used for the experimental results on the Bayesian direct filter reported in Section 6.4. The QinetiQ radar data were collected at a coastal site in the United Kingdom; this second database was used for the experimental results reported in reference 10.

6.7.1 Target-to-Interference Ratio

For meaningful comparison among different receiver structures, target detection performance was parameterized by the power of the target signal relative to the background of all other interfering signals. For the study at hand, the interfering signals include clutter and front-end receiver noise. The parameter is called the *target-to-interference ratio* (TIR), which is measured in decibels.

From inspection of real-life data, it was evident that (e.g., within the time of one Doppler spectrum) the instantaneous power of both the target and clutter vary significantly. Specifying a single representative TIR therefore becomes particularly rather difficult.

From an experimental viewpoint, one of the attractions of creating a "synthetic" target signal is that all its properties, including power, are known and therefore under experimental control. In the injection of synthetic targets described in reference 12, the TIR is evaluated from the target and clutter/noise data on a per-Doppler-spectrum basis, since this is the parameter from which the pertinent Bayesian probabilities are calculated. Accordingly, for convenience of presentation, the amplitude of the synthetic target signal was kept constant, with the natural variation of the clutter power providing a spread of TIR values within a given data file. To provide the effect of a target gradually appearing and disappearing, the target amplitude is *modulated* on a long-term-basis; an example of this provision is shown later.

Figure 6.9 shows the time-Doppler image of a typical target-plus-interference signal; the figure is an example of a synthetic target injected into the IPIX clutter data using VV polarization.

6.7.2 Receiver Comparison

This section compares the probability of detection, P_d, and probability of false alarm, P_{fa}, produced by the two receivers when operating on the same dataset.

Unfortunately, due to the differing nature and data requirements of the two receivers, it is not possible to make an "apple-to-apple" comparison. For example, based on the experience with use of the Bayesian receiver on IPIX data, we need to use a frame size of 64 or 128, along with a forward-backward window size of 33, in order to make a target declaration. This represents the use of $64 * 33 = 2112$ or $128 * 33 = 4224$ data samples, respectively. The Bayesian receiver needs at least this many number of samples in order to perform well. In contrast, the correlation

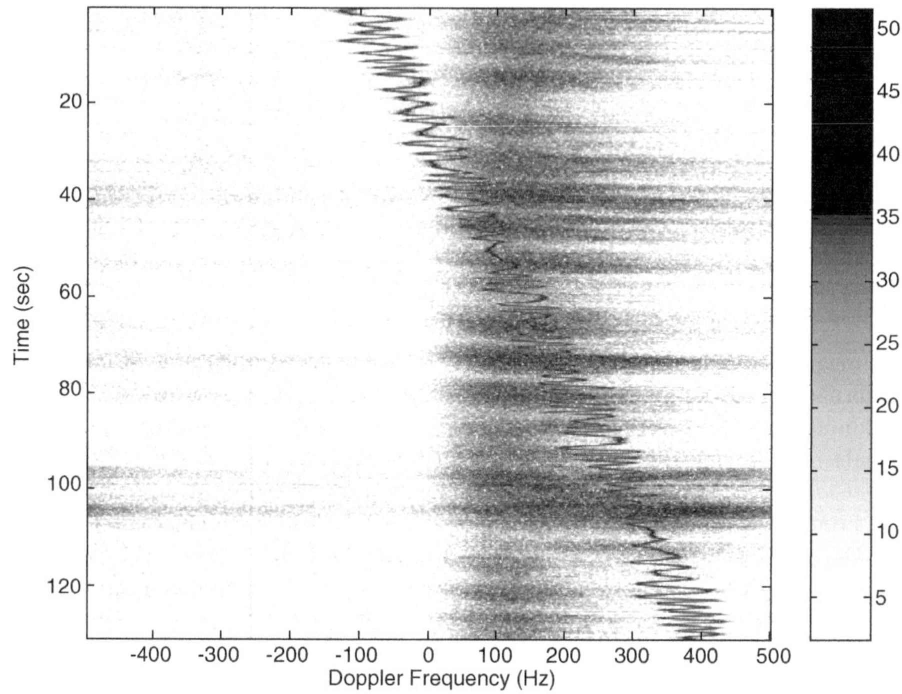

Figure 6.9 Example of a synthetic target injected into real IPIX clutter data, VV polarization.

anomaly receiver processes 1000 samples at a time. This latter receiver is based on the principle that the *K*-distributed signal is Rayleigh-distributed over short periods of time. Increasing the number of data points (hence the time period) beyond 1000 (1 s at a PRF of 1000 Hz) would make this assumption increasingly less valid, thereby reducing receiver performance. Therefore, we compare the performance of the two receivers using the same dataset, but with some difference in the number of samples that each receiver uses in deriving its detection decision.

The first comparison was done using the QinetiQ HH polarization data (Fig. 6.10). The results for the correlation anomaly receiver are shown in Fig. 6.11. The Bayesian receiver was operated, in a manner similar to that described earlier for the IPIX data, at each successive range index in the range swath. A target was declared if the maximum probability of a target's being present occurred in a Doppler bin other than the bin used for "no target present." Figure 6.12 shows the result for the Bayesian receiver processing QinetiQ HH data using a frame size of 64, a forward–backward window size of 33, and a target inertia parameter of 3.0. Although the target is visible, there are numerous false alarms. Increasing the frame size to 128 retains the target response but removes nearly all of the false alarms, as shown in Fig. 6.13.

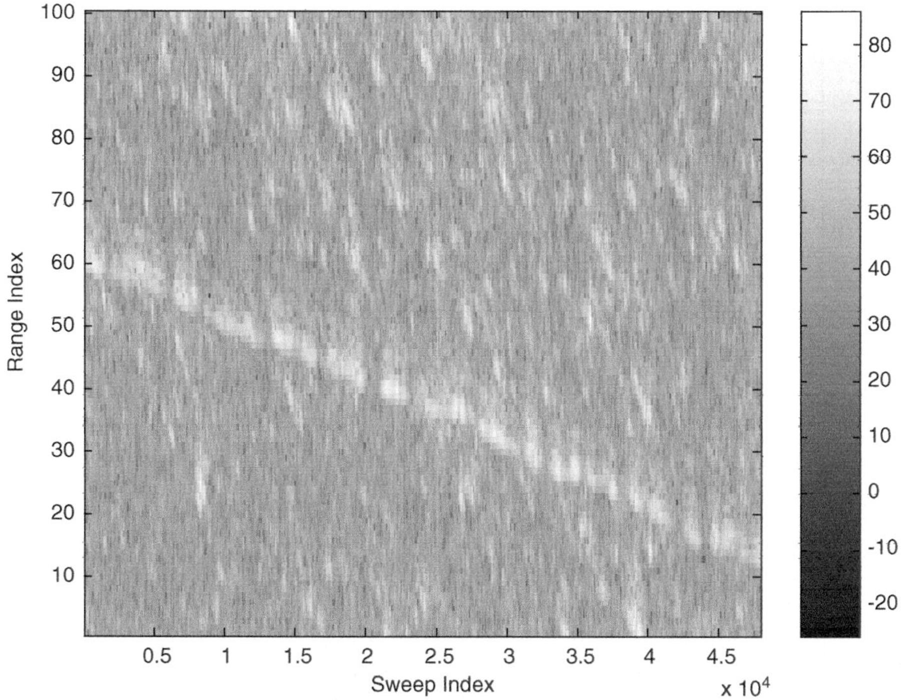

Figure 6.10 Intensity (arbitrary decibel scale) of a swath of the QinetiQ HH "coke can" data, 100 ranges by 48,000 sweeps. PRF is 1000 Hz. The target (can) is apparent, closing in range from index 61 (approximate) to index 13 (approximate) over the 48-s duration of the file. In terms of the complete data file, this swath starts at range 751.

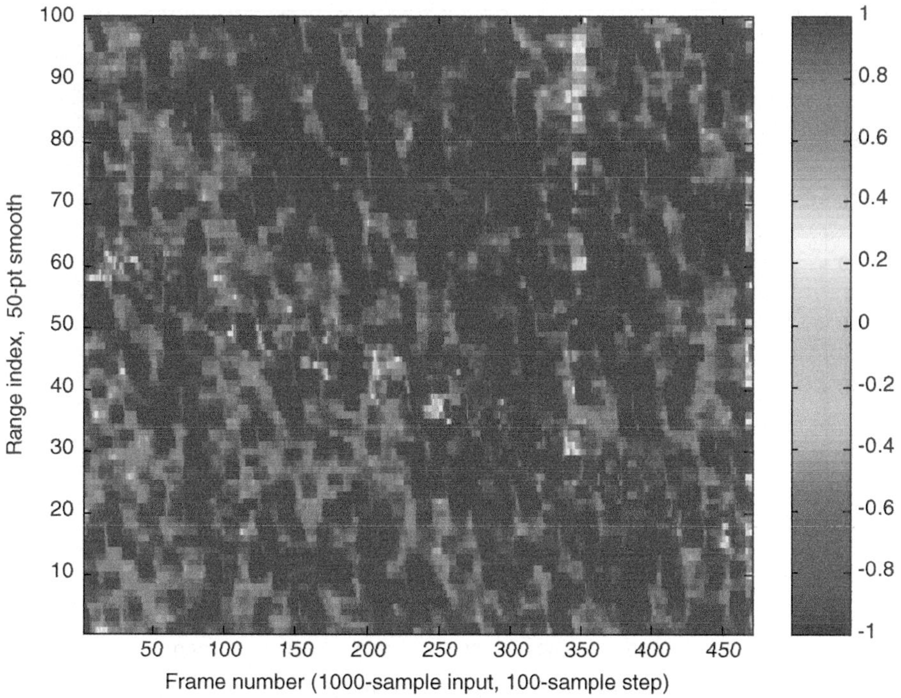

Figure 6.11 Image of 1-correlation for correlation anomaly receiver applied to the HH data shown in Figure 6.10. The effect of the target is faintly visible for some of the time. There are also two vertical anomalies, around frame numbers 350 and 470. These can be traced to data glitches (i.e., erroneous sweeps).

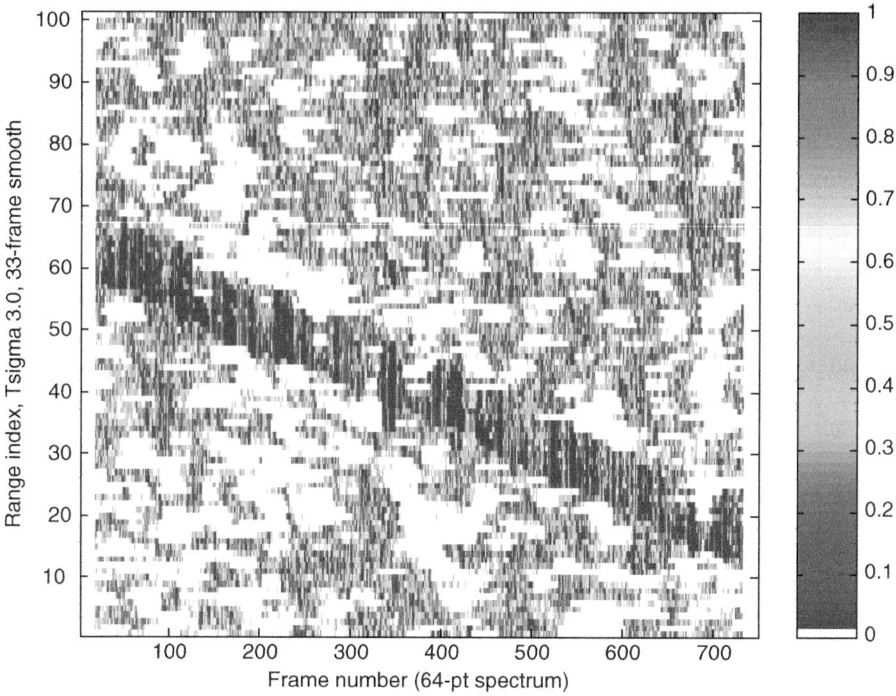

Figure 6.12 Result of processing the data of Figure 6.10 using the Bayesian receiver with frame size 64, forward–backward window size of 33, and target inertia parameter of 3.0. The response to the target is clearly visible, but there are numerous false alarms.

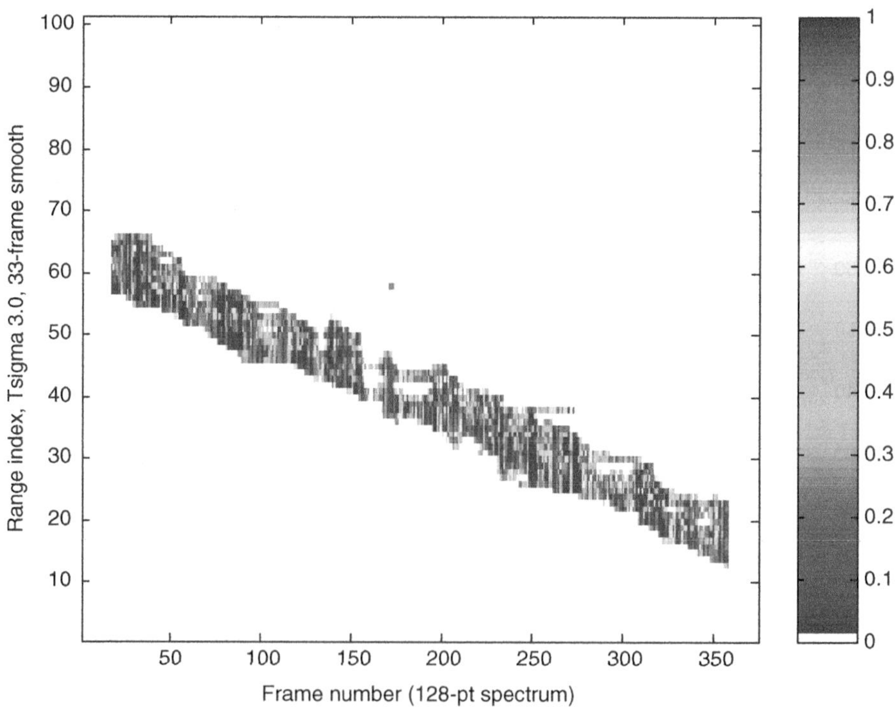

Figure 6.13 Result of processing the data of Figure 6.10 using the Bayesian receiver with frame size 128, forward–backward window size of 33, and target inertia parameter of 3.0. The response to the target is readily detected, and there are almost no false alarms. The target has an extended return in range due to pulse compression sidelobes and/or digital sampling effects.

To provide a more quantitative receiver comparison, the correlation anomaly receiver was tested using simulated targets in the same IPIX clutter as used to test the Bayesian receiver earlier. The detection decision metric for the correlation anomaly receiver is the value of the correlation, c_ψ. If it falls below a given threshold, an anomaly (i.e., target) is declared. First, the cumulative distribution function of the correlation value for interference-only data was estimated, shown in Figure 6.14, and used to select the value for the detection threshold. Then the correlation anomaly receiver operated on a simulated target plus clutter, on a range cell by range cell basis. At each range, a time window of 1000 samples was used to estimate a single value of c_ψ, then the start of the next window was stepped forward in time by 50 samples. This produced about 2600 estimates of c_ψ from the 131 s of data per range cell. A target detection was declared when the correlation fell below the detection threshold. A two-dimensional image of P_d, similar in form to that for the Bayesian receiver, was produced, as shown in Fig. 6.15. It exhibits a very unusual behavior. Reliable detection, regardless of the TIR, only occurs when the mean target Doppler frequency is much less than the mean frequency of the clutter. Figure 6.16 shows the correlation versus time for the time history of a synthetic target in clutter at a single-range cell. The correlation shows the effect of the target only for about the first 25% of the 131 s of time history. With reference to Fig. 6.9, the target shows detectability when its Doppler frequency is negative, while the clutter Doppler

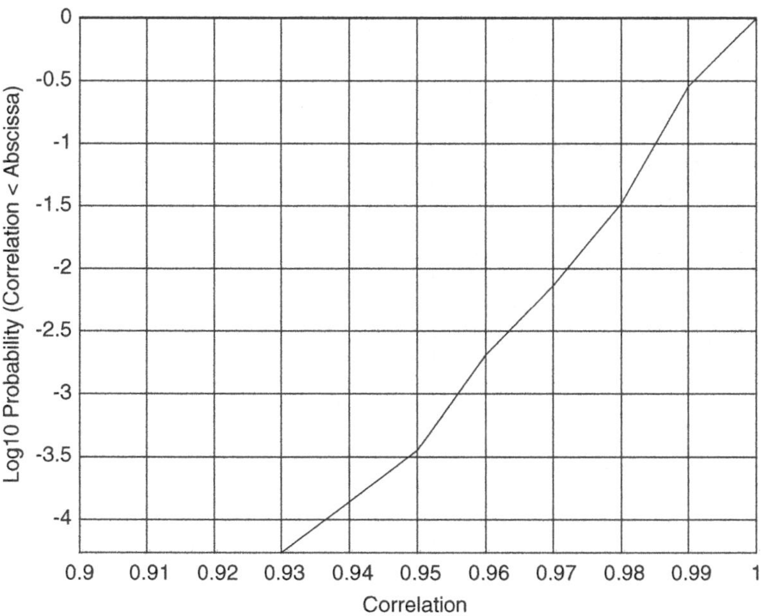

Figure 6.14 Probability that correlation for interference-only will fall below a certain value for the correlation anomaly receiver. For a given threshold value for the correlation, an estimate of the P_{fa} can be made. For example, using a target detection threshold of 0.95 yields a P_{fa} of about $10^{-3.5} = 3*10^{-4}$.

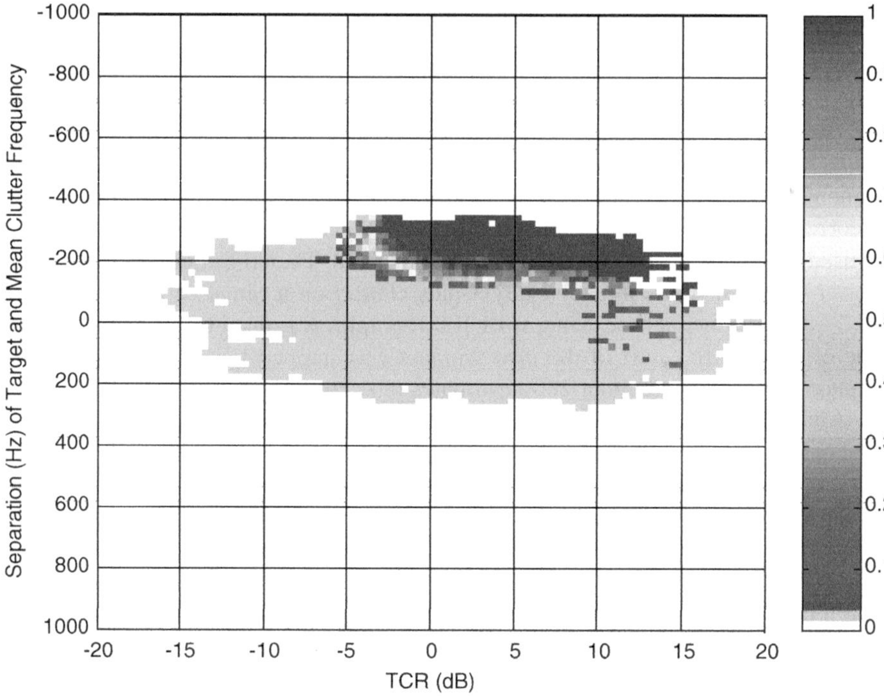

Figure 6.15 Probability of detection performance for the correlation anomaly receiver operating on a simulated target in HH IPIX data. Gray color indicates where the opportunity for at least one detection occurred, but the P_d was zero. A target detection was declared when the correlation fell below the detection threshold of 0.95. Note that the receiver only provides detections when the target mean frequency is less than the mean clutter frequency by at least 200 Hz.

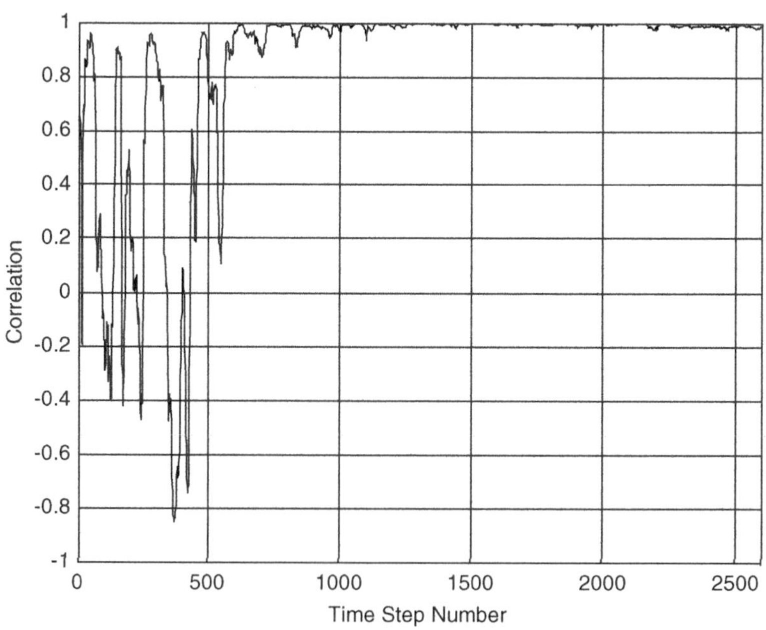

Figure 6.16 Correlation versus time for the time history at a single-range, HH polarization, for the correlation anomaly receiver.

frequency content is always positive. The reason for this behavior has not yet been determined.

To test that this result is not peculiar to the particular target behavior, the translational frequency behavior of the target was changed to that shown in Fig, 6.17. Figure 6.18 shows the corresponding correlation behavior. Again, the receiver seems to respond to the presence of the target only when the target and clutter frequency contents are in separate halves of the Doppler spectrum (negative versus positive frequency halves).

To further test this behavior, another IPIX file was used, one in which the clutter occupied only negative frequencies. Figure 6.19 shows the simulated target in this new clutter scenario. Figure 6.20 shows the interference-only correlation cumulative distribution function. For this file, the response is more widespread, requiring a threshold of 0.4 to achieve a P_{fa} of 10^{-3}. Figure 6.21 shows the final P_d performance. This time the target is detectable only when its frequency is above the mean clutter frequency. Figure 6.22 confirms this behavior, with the correlation showing the effect of the target only when the target frequency is positive; the clutter mean frequency is always negative.

Figure 6.23 shows the detection performance for the Bayesian receiver for vertical–vertical (VV). To produce no measurable false alarms, the required parameters

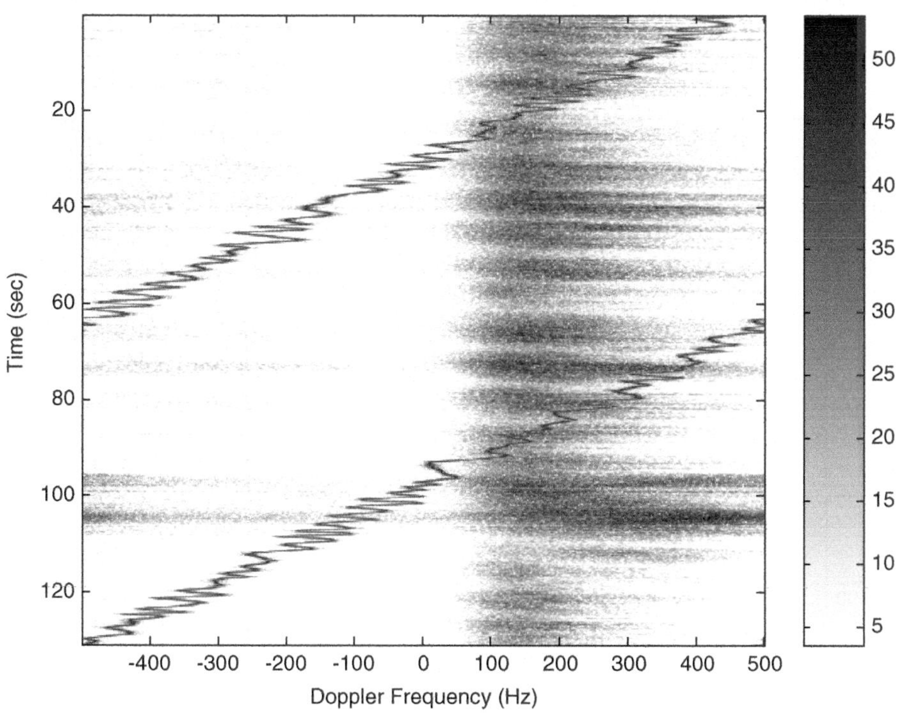

Figure 6.17 Test data used to confirm behavior of correlation versus separation of target and clutter mean frequencies.

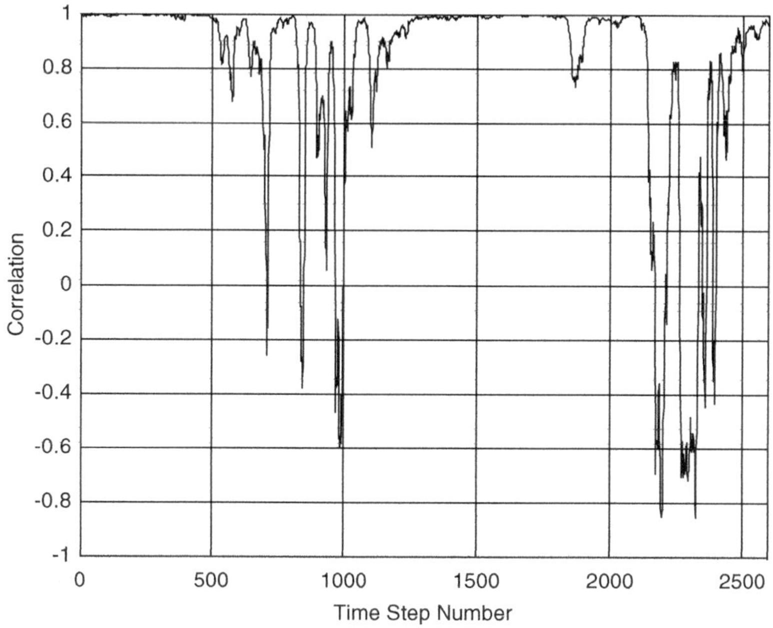

Figure 6.18 Correlation for the anomaly receiver for data shown in Fig. 6.17. Target seems to generate a reduced correlation only when the target and clutter frequency content is contained in opposite halves of the Doppler spectrum.

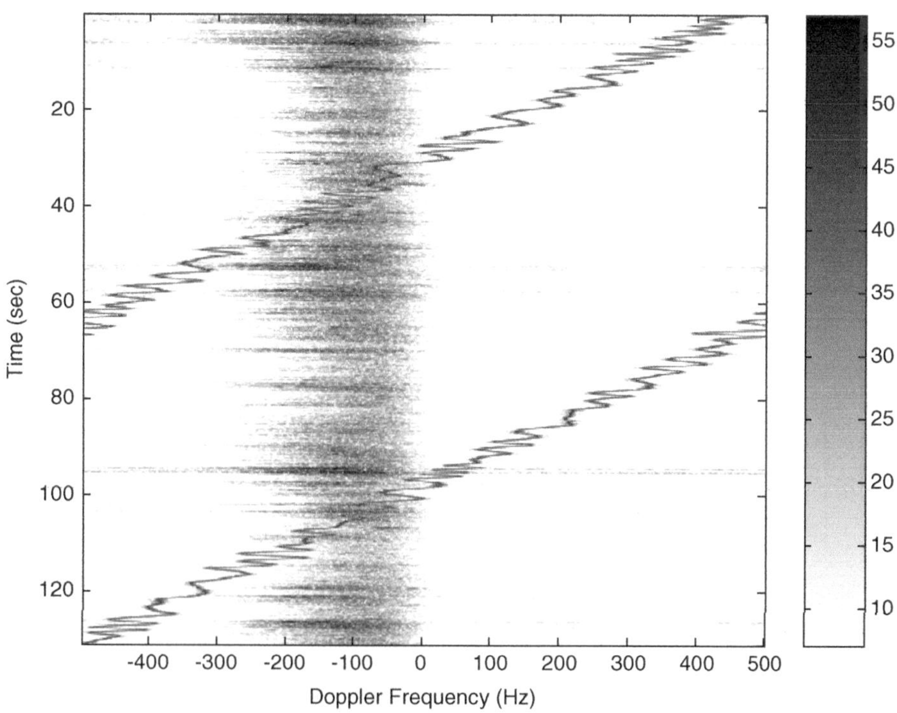

Figure 6.19 Different IPIX clutter file for which clutter exhibits only negative Doppler frequencies.

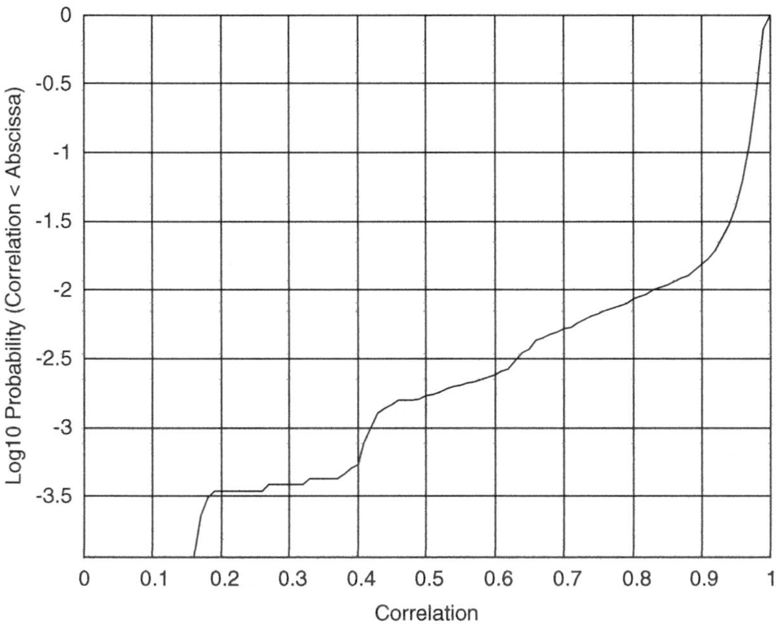

Figure 6.20 Probability for the anomaly receiver that correlation for interference-only will fall below a certain value, for the IPIX data file as in Fig. 6.19.

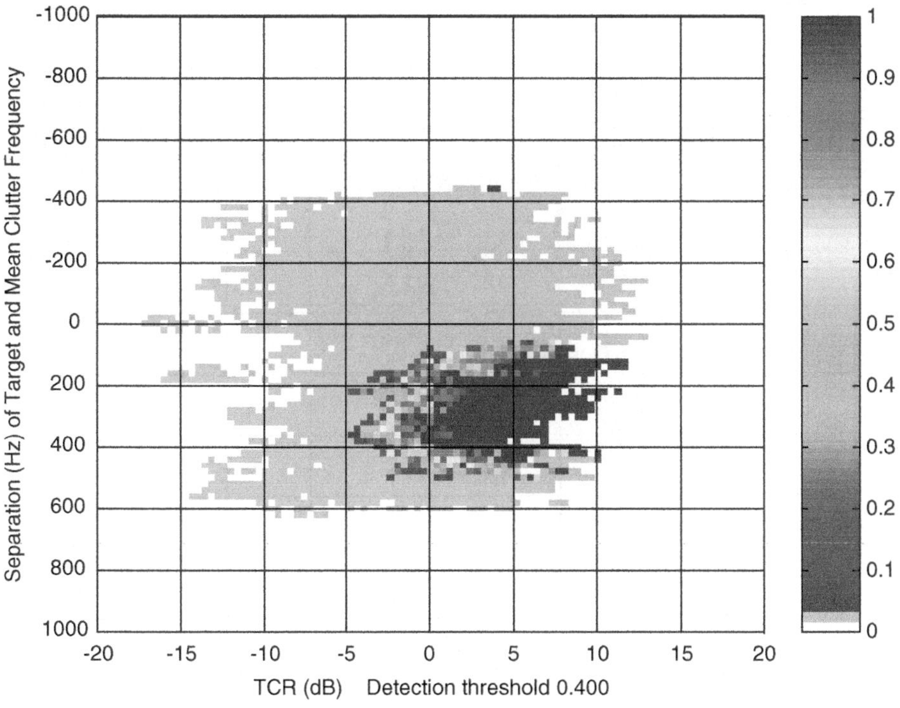

Figure 6.21 P_d performance for the correlation anomaly receiver for simulated targets in clutter as in Fig. 6.19. Target effect is only detectable if target frequency is above clutter frequency. Note that a frequency separation greater than 500 Hz actually spectrally aliases to negative frequencies in the Doppler spectrum.

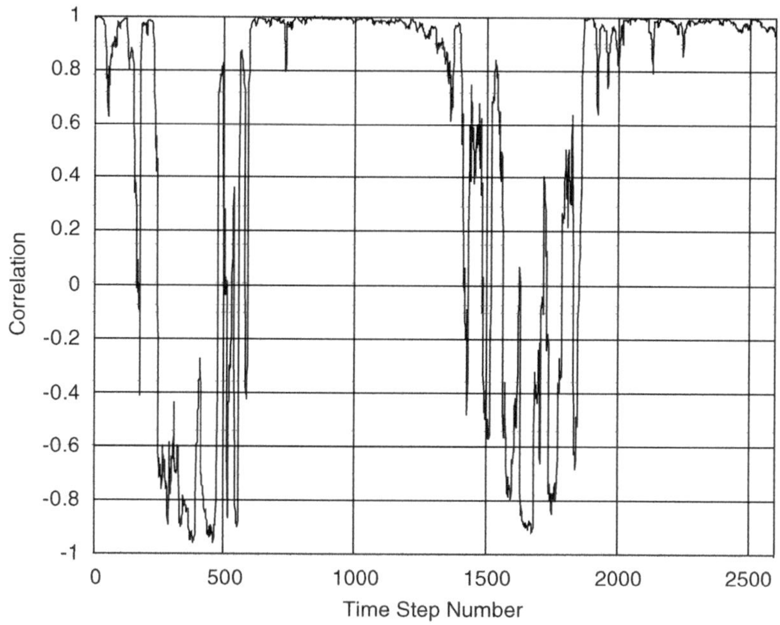

Figure 6.22 Correlation for the anomaly receiver for data shown in Fig. 6.19.

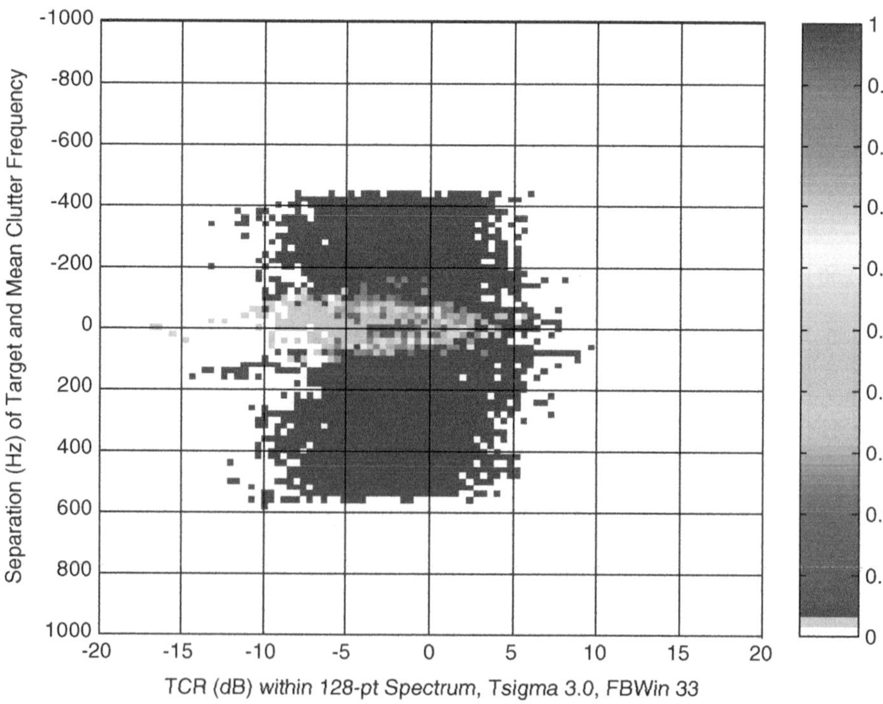

Figure 6.23 P_d performance for the Bayesian receiver operating on VV IPIX data, similar to that shown in Fig. 6.19. The frame size was 128, the forward–backward window size was 33, and the target inertia parameter was 3.0. The gray pixels are locations for which at least one detection opportunity was available, but the P_d was zero.

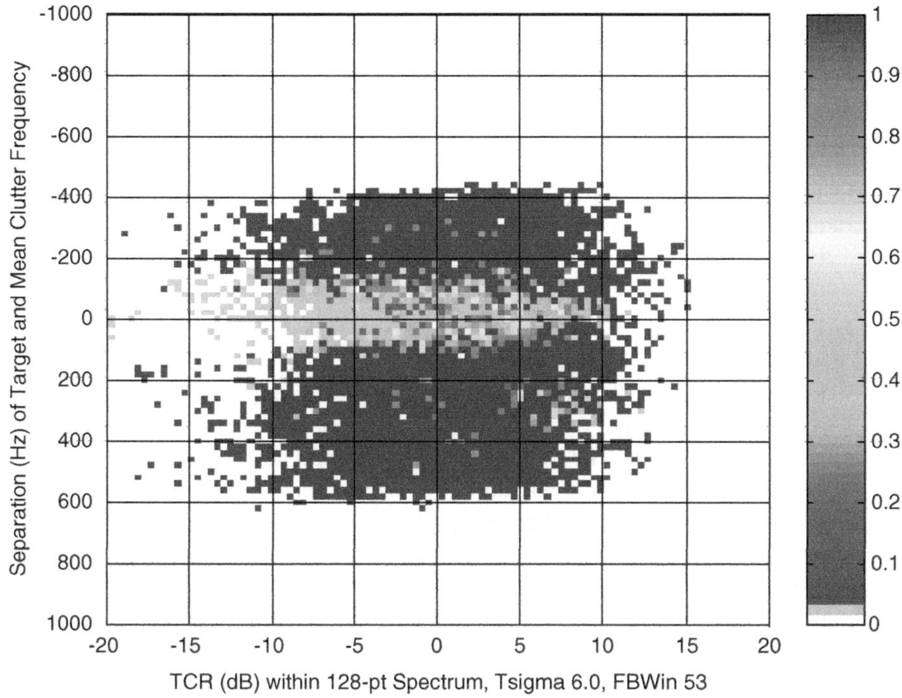

Figure 6.24 P_d performance for the Bayesian receiver operating on HH IPIX data, similar to that shown in Fig. 6.19. The frame size was 128, the forward-backward window size was 53, and the target inertia parameter was 6.0. The gray pixels are locations for which at least one detection opportunity was available, but the P_d was zero.

were a frame size of 128, forward–backward window size of 33, and target inertia parameter of 3.0. Figure 6.24 shows the P_d result for HH data. To maintain no measurable false alarms, the frame size was 128, but the forward–backward window size was increased to 53, and the target inertia parameter was increased to 6.0. For both HH and VV data, the Bayesian detector provides usable detection performance at most target frequencies, but requires higher target-to-interference ratios when the target and clutter frequencies overlap.

6.8 DISCUSSION

In this chapter, we described two new radar detection strategies:

- The first strategy, referred to as a direct filtering algorithm, is rooted in the Bayesian framework.
- The second strategy, referred to as the correlation anomaly detection algorithm, is rooted in stochastic differential equation (SDE) theory.

We also presented a comparative evaluation of these two algorithms using two different datasets, which included the creation of a "synthetic" target signal, thereby providing a means for experimental control. Although it was not possible to do a proper "apples with apples" comparison of these two new algorithms because the two algorithms naturally use different amounts of data, an attempt was made to compare them as fairly as possible.

Here is a summary of the results of this comparison:

1. Using the QinetiQ data, Fig. 6.11 shows the processor output for a target in clutter for the correlation receiver. Comparing this with Fig. 6.13 for the Bayesian receiver output qualitatively, we see that the Bayesian receiver makes the target much more visible.
2. For a more quantitative comparison, we used real IPIX clutter data with injected simulated targets. Using data of the nature depicted in Fig. 6.19, Fig. 6.21 shows the detection performance of the correlation receiver, while Fig. 6.23 shows the corresponding performance of the Bayesian receiver. The figures display the probability of detection as a function of target-to-clutter ratio (x-axis) and the separation in Doppler of the target and mean clutter frequencies (y-axis). Clearly, the Bayesian receiver has a more robust performance than the correlation anomaly receiver. There remains the issue of why the correlation receiver is sensitive to the target-clutter Doppler frequency separation. This phenomenon remains unexplained.

6.8.1 Further Research

In describing both the Bayesian receiver and the correlation anomally receiver, we focused on the detection of a single target in sea clutter. In a typical radar surveillance environment, however, the radar has to deal with *multiple targets*. Extension of both of these algorithms to deal with this more general situation is an issue that needs to be studied.

In the context of the Bayesian receiver, we conclude the discussion with one last comment. When dealing with unknown targets in clutter, the tracking capability of this receiver may be strengthened by incorporating in algorithmic formulation of the receiver, a *variable rate state* defined in terms of a vector $\boldsymbol{\theta}_t$ and state arrival time τ_t, where, as before, t denotes discrete time. The vector $\boldsymbol{\theta}_t$ parameterizes the target state by such variables as position, velocity, and acceleration. For analytic purposes, the variable rate state could be assumed to follow a Markovian model. In so doing, we may be able to formulate a new tracking algorithm termed the *variable rate Bayesian receiver*, building on ideas described in the paper by Godsill and Vermaak [13] on variable rate particle filters; particle filters are also referred to in the literature as sequential Monte Carlo methods. In recent years, they have become popular for tracking applications made difficult by having to deal with nonlinear and nonGaussian state-space models, which are frequently encountered in real-life environments.

REFERENCES

1. Y. Bar-Shalom and A. Jaffer (1972). Adaptive nonlinear filtering for tracking with measurements of uncertain origin, in *Proceedings of the 1972 IEEE Conference on Decision and Control, New Orleans, LA, December 1972*, pp. 243–247.
2. Y. Bar-Shalom and E. Tse (1975). Tracking in a cluttered environment with probabilistic data association. *Automatica* **11**, 451–460.
3. D. B. Reid (1979). An algorithm for tracking multiple targets. *IEEE Trans.* **AC-24**, 843–854.
4. X. Rong Li (1998). Tracking in clutter with strongest neighbor measurements—part i: Theoretical analysis, *IEEE Trans. Automatic Control* **43**, 1560–1568.
5. M. G. S. Bruno and J. M. F. Moura (2001). Multiframe detector/tracker: Optimal performance. *IEEE Trans. Aerosp. Electron. Syst.* **37**, 925–944.
6. S. Haykin, R. Bakker, and B. Currie (2002). Uncovering nonlinear dynamics: The case study of sea clutter, *Proc. IEEE* **90**(5), 860–881.
7. R. Bakker, B. Currie, T. Kirabarajan, and S. Haykin. Adaptive radar detection: A Bayesian approach, EPSRC and IEE Workshop on Nonlinear and Non-Gaussian Signal Processing, Peebles, Scotland, July 8–12, 2002.
8. R. Bakker, G. Lopez-Risueno, and S. Haykin, Bayesian approach to the direct filtering of radar targets in clutter, August 2002 (Internal Report).
9. T. Field, Diffusion processes in electromagnetic scattering generating *K*-distributed noise, EPSRC and IEE Workshop on Nonlinear and Non-Gaussian Signal Processing, Peebles, Scotland, July 8–12, 2002.
10. T. R. Field and R. J. A. Tough (2003). Diffusion processes in electromagnetic scattering generating *K*-distributed noise, *Proc. R. Soc.* London **A459**, 2169–2193.
11. M. B. Priestley (1981). *Spectral Analysis and Time Series.* Academic Press, New York.
12. B. Currie and S. Haykin, Bayesian detector evaluation and comparison, Final Report, Adaptive Systems Laboratory, McMaster University, April 2004. (Report prepared for Defence Research and Development Canada-Ottawa, under Contract No. W7714-02683/001/SV.)
13. S. J. Godsill and J. Vermaac (2005). Variable rate particle filters for tracking applications. In *Proc. IEEE Statistical Signal Processing Workshop*, Bordeaux, France, 6 pages.

Index

Page references followed by t indicate material in tables.

E

F

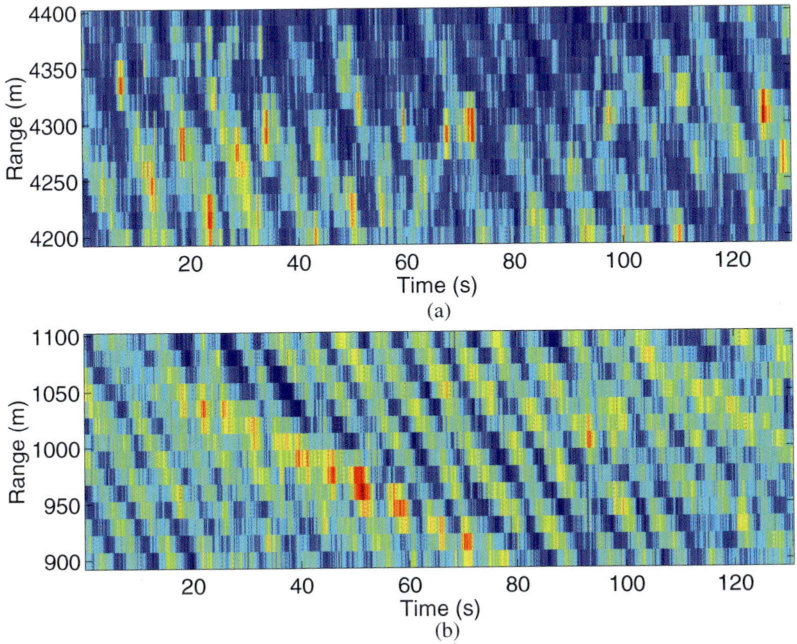

Figure 4.5 Radar return versus time and range, VV polarization, of (a) dataset $L2$ (low sea state) and (b) dataset H (high sea state). The color axis shows $\log(|\tilde{x}|)$, where $\tilde{x}$ is the complex envelope of the received signal. The units of $\tilde{x}$ are normalized and the color axis changes from blue (low) via green to red (high).

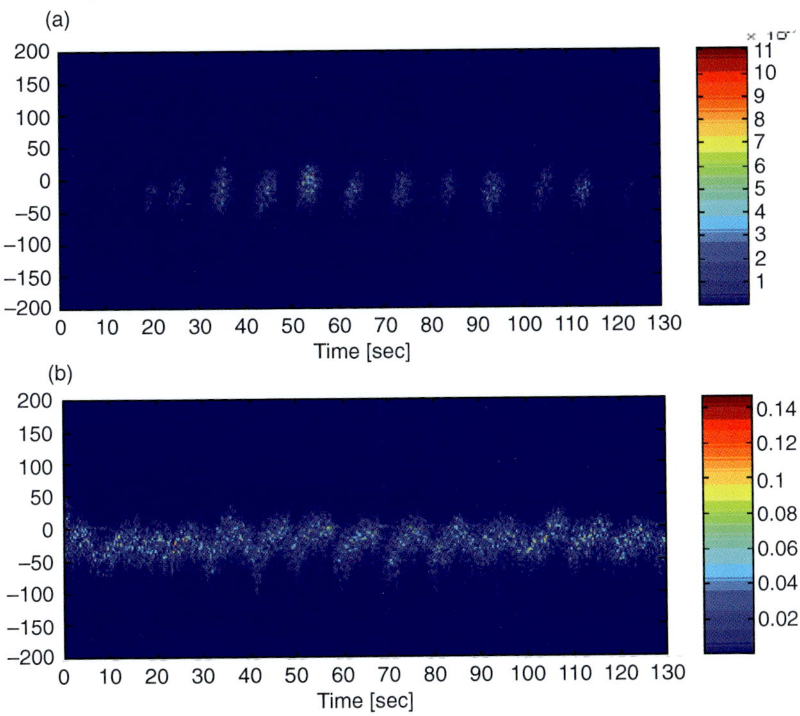

Figure 5.8 (a, b) VV spectrogram and normalized spectrogram. Dataset: Starea4 of November 7, range cell 3.

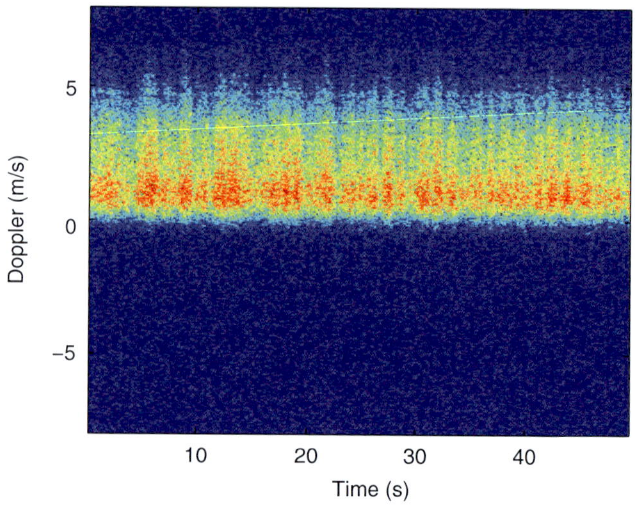

Figure 4.10 Time-Doppler spectrum of 50 s of data synthesized with the method of Conte, Longo and Lops [43]. (Data provided by Alan Thomson, DRDC, Ottawa.)

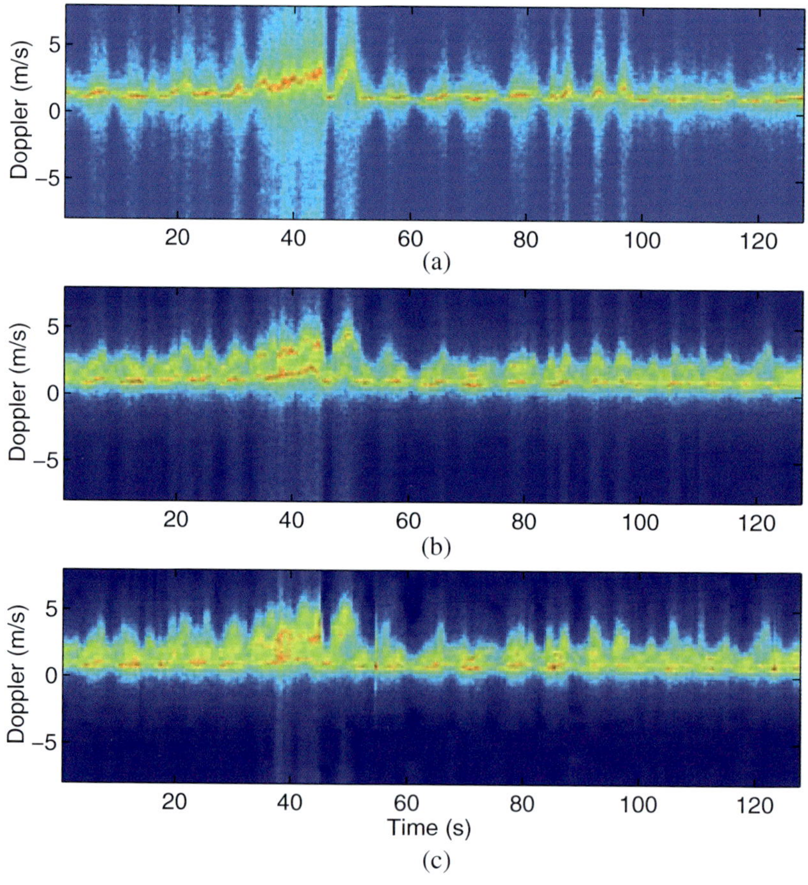

Figure 4.13 Time-Doppler spectra of data synthesized by a sliding AR process of order 1 (a), order 2 (b), and order 3 (c). The three plots of the figure have identical color axis limits. The lighter background color in plot (a) is caused by the larger residual error of the sliding AR(1) model.